SOUND AND ACTION IN MUSIC PERFORMANCE

SOUND AND ACTION IN MUSIC PERFORMANCE

Peter Q. Pfordresher
Department of Psychology, University at Buffalo, The State University of New York Buffalo, NY, United States

ELSEVIER
ACADEMIC PRESS
An imprint of Elsevier

Academic Press is an imprint of Elsevier
125 London Wall, London EC2Y 5AS, United Kingdom
525 B Street, Suite 1650, San Diego, CA 92101, United States
50 Hampshire Street, 5th Floor, Cambridge, MA 02139, United States
The Boulevard, Langford Lane, Kidlington, Oxford OX5 1GB, United Kingdom

British Library Cataloguing-in-Publication Data
A catalogue record for this book is available from the British Library

Library of Congress Cataloging-in-Publication Data
A catalog record for this book is available from the Library of Congress

ISBN: 978-0-12-809196-8

Publisher: Nikki P. Levy
Acquisition Editor: Emily Ekle
Editorial Project Manager: Barbara Makinster
Production Project Manager: Mohanapriyan Rajendran
Cover Designer: Greg Harris

Typeset by MPS Limited, Chennai, India

Contents

Preface

This book concerns a very specific scientific question that has far-reaching implications: How do music performers rely on auditory perception of the sounds that they themselves create? We call these sounds *auditory feedback*. This focused question, which has dominated my own research for the past 15 years, brings in many core questions from the psychology of human performance, including the way perception and action interact, how people plan and execute complex event sequences, the perception of auditory events, rhythmic coordination of actions, differences between music and language, the representation and perception of time, and sensorimotor learning. I discuss behavioral and neuroscience research, and in so doing make connections across research to propose a hypothetical model of perception and action in music called the *Hierarchical Perception–Action Coordination* model, or H-PAC. This model is most explicitly developed in Chapters 1, 3–5, and 8.

Interest in how perception and action interact has increased greatly, both within music cognition and in cognitive science generally speaking. In drafting this volume I have attempted to avoid overlap with existing noteworthy contributions to this area. For instance, those interested in how action influences perception (the converse of my focus in this book) may wish to consult Leman (2008), those interested in a more musicological treatment of themes in this book may wish to consult De Souza (2017), and those more interested in the intersection of perception and action for speech may wish to consult Guenther (2016).

The first chapter of this book provides an introduction to the general issues under consideration, the existing relevant theories, and experimental methodology. Readers already familiar with this area of research could skim the first chapter. Chapters 2–5 then summarize the core research on the role of auditory feedback in music performance, developing the aforementioned theory in the process. Different chapters focus on different kinds of experimental effects that have distinct theoretical implications. The first five chapters thus constitute the core of this book. The remaining chapters then delve into more specialized topics of interest that may qualify the implications of the general findings discussed in Chapters 1–5. Each chapter integrates behavioral and cognitive neuroscience results, with results from the latter area usually discussed in a separate section.

I have written this book in a way that I hope can be accessible to a broad range of readers, with the expectation that most readers will probably have some degree of specialization in music cognition and cognitive neuroscience. I have attempted to limit the level of music theory background and background in statistics and neuroscience that one needs for this book, and have adopted a more conversational writing style than I would use in a journal article.

Many outside factors helped me in the process of writing this book. Starting from the beginning, I would like to thank my mentors during graduate school for their significant influence in directing me toward this area of research. Peter Howell first pointed me to the DAF effect during my master's degree, although at the time ironically I pursued a different project. Caroline Palmer later suggested using altered feedback manipulations for my dissertation, and Mari Riess Jones gave me a theoretical background that inspired to a large degree how I think about these results. I am indebted both to the Fulbright Canada Foundation and the University at Buffalo for providing me with time away from teaching and service (a fellowship from the former coupled with a sabbatical from the latter) during the first half of 2018. Moreover, the Fulbright award led me to the extensive music neuroscience community in Montreal. While there, my knowledge of the research broadened greatly, and I was able to receive extensive feedback on book chapters from Caroline Palmer, Anna Zamm, and Rebecca Scheurich. I also received valuable feedback from Steve Finney and Lyn Pfordresher. I also want to thank my family: Lyn, Emma, and Paul, for their support and patience with me as I spent long hours upstairs working on the book.

To close this brief preface, I wish to dedicate this book to the memory of my father-in-law, William A. Canterbury, who tragically passed away while I was working on this book. He was an excellent writer—a former journalist at the *Akron Beacon Journal*—and a loving family man who we lost too soon.

Thank you for taking the time to read this work. I hope the research I talk about here fascinates you as it has me.

Peter Q. Pfordresher
Buffalo, NY, United States
August 2018

CHAPTER

1

Setting the Stage

When the Beatles played concerts to crowds of screaming teenagers, they could barely hear themselves, and yet they sung in three-part harmony with astounding accuracy. On the other hand, speaking into a telephone with a delay in the sound can be tremendously disruptive. About 10 years ago, my university installed fancy "IP-drop" phones in all the offices. Although I enjoyed the sleek new design, my new phone had and still has a problematic quirk. Occasionally, while I'm talking, I hear a copy of my own voice that occurs a fraction of a second after I speak (this can occur when the microphone on the recipient's phone picks up the output from your phone). The effect of this delay is maddening. When it happens, I become much more selfconscious of my voice. It becomes hard to pace my speech appropriately and find myself occasionally stuttering and drawling out my vowels, features of speech referred to as *disfluencies*. I find it impossible to ignore this rebroadcast of myself. Usually speaking at a slower rate helps a bit, but I almost always have to apologize to my interlocutor, who is unaware of the glitch. A roughly analogous effect used to occur for international calls, as there used to be a considerable delay in the time at which your phone would receive the signal from your interlocutor. In a twist of irony, I have used exactly this sort of manipulation on research participants for the past 20 years.

These examples concern the *altered auditory feedback* effect, a paradigm that will dominate the research discussed in this book. The alteration described above involves an alteration of timing—a delay in the onset of speech feedback relative to when speech is produced. However, alterations can be made to any feature of sound. Using modern digital technology, my voice could be altered to sound like another person's, could be shifted in pitch, or could be made to sound like different words. Although the example I just gave has to do with speech, similar effects are found in music performance, which is the main focus of this book.

Sound and Action in Music Performance
DOI: https://doi.org/10.1016/B978-0-12-809196-8.00001-2

The effects of altered feedback can be entertaining or frustrating, but they also shed light on an important question: What role does perception have in performance? When you perform music, you have to plan, execute, and control a complex sequence of actions that can involve a sequence of finger movements (e.g., piano), or changing states of muscular tension (e.g., singing). The demands on the motor system, in the brain, muscles, and joints, are considerable. However, actions do not really constitute the *goal* of a musical performance. You perform music to create a pattern of *sound* that your listeners can perceive and enjoy. Thus, action planning in music performance may result from planning an intended *sound pattern*.

This book will address how the sound of one's own performance (i.e., auditory feedback) influences the continued planning and the execution of music. Although my discussion will be dominated by research on the effects of altered auditory feedback, I will consider other paradigms that have to do with the role of sound in music performance. For instance, in Chapter 9 I discuss research on how performers use auditory information from each other to synchronize their performances. I will also consider the process by which people translate a melody into a sequence of actions when they play or sing "by ear," a form of imitation. Although my main focus is on music, I will also consider the role of auditory feedback in speech (Chapter 7). Speaking and music making are the primary way people communicate their thoughts and feelings through the auditory modality, and thus auditory feedback is critical in each case.

For the most part, my discussions of music performance will concern the piano, with somewhat less focus on vocal performance, and unfortunately, very little on other kinds of instrumental performance. This limitation reflects the scope of music cognition research on performance. Piano outweighs other instruments by far in its popularity, and electronic pianos allow researchers to measure all relevant dimensions of performance with high efficiency and precision in the lab. Furthermore, the flexible mapping of actions to sound that is available on the piano make it an interesting test case for sensorimotor learning (as discussed further in Chapter 6). Although singing includes an even larger portion of the population (potentially everybody), the difficulty of measuring vocal production for some time inhibited its use in empirical studies. Another limitation of studying the voice, as I will discuss in Chapter 7, is the difficulty of implementing certain alterations to auditory feedback. Nevertheless, an increasing amount of research on singing has emerged, which I will discuss, which is fascinating in part due to the presence of large individual differences in pitch accuracy of singers (Berkowska & Dalla Bella, 2009, 2013; Dalla Bella, Giguère, & Peretz, 2007; Hutchins & Peretz, 2012a, 2012b; Pfordresher & Brown, 2007; Pfordresher & Larrouy-Maestri, 2015).

In this introduction, I lay out the basic territory that this book explores. I start by discussing the production of action sequences, focusing on music performance. I then discuss what we mean by perceptual "feedback" and consider some nuances associated with this term. Finally, I provide an overview of some of the leading theoretical and empirical treatments of perceptual feedback.

INTENTIONS, PLANNING, AND EXECUTION IN MUSIC PERFORMANCE

A great deal of psychological research considers discrete actions (e.g., a key press) or actions oriented toward a single goal (e.g., grasping a dowel). Such behaviors are highly useful given the amount of experimental control they allow, and the easy way that one can separate planning from execution. However, most of the time we plan and execute behaviors in a sequence. Even when grasping an object (e.g., a kettle) we are often doing so when preparing a subsequent goal (pouring into a coffee filter). As such, it is important for research on performance to consider the constraints on sequential actions.

Music is a quintessentially sequential behavior. A single note or chord means almost nothing outside of a context. An "F" may be a stable ending pitch in a happy tune (F-major), or it may cause a melody to sound sorrowful (D-minor), or may sound out of place (B-major). The sequence defines the meaning. A similar case can be made for speech. Single words can convey meaning ("Fire!"..."Yum!"). However, as in music the meaning of speech usually comes from sequential context ("The candidate lacks *fire* in the belly," "I'm on *fire* with love for you," or "Help, help...*FIRE!*"). Of course, there are many ways in which music performance differs from speech. I consider the similarities and differences between these two forms of expression in Chapter 7.

A particularly important component of music performance is the formation of an *action plan*: a series of motor commands based on an intended outcome. Let us now consider the link from intention to plan to execution. In so doing I will build the framework for a theory of perception and action in music that will be developed throughout this book: the Hierarchical Perception/Action Coordination (H-PAC) model.

First, what is a musical intention? Intentions are fundamental to psychology, and in many areas of behavior the notion of an intention is incredibly complex. What has a child intended when she throws a valuable china bowl on the floor? We have a somewhat simpler situation in music, though not without its own intricacies. For now I propose that a performer's intention is an *auditory image of a correct performance*. In other words, musicians intend to create sound sequences, they do not directly intend to produce movements. Of course, there are some situations in which a performer may not have formed an auditory representation of the piece they want to play. This can happen if someone is reading a piece for the first time from notation, and the individual lacks the kind of reading skills that lead to associations of notation with auditory imagery (called notational audiation, Gordon, 1975). In such cases I would argue that the performer has not truly formed an intention for the piece of music, but instead is following a series of scripted commands. In other words, the notation in these cases stands as a kind of proxy for the intention. In Fig. 1.1, an intention is represented as

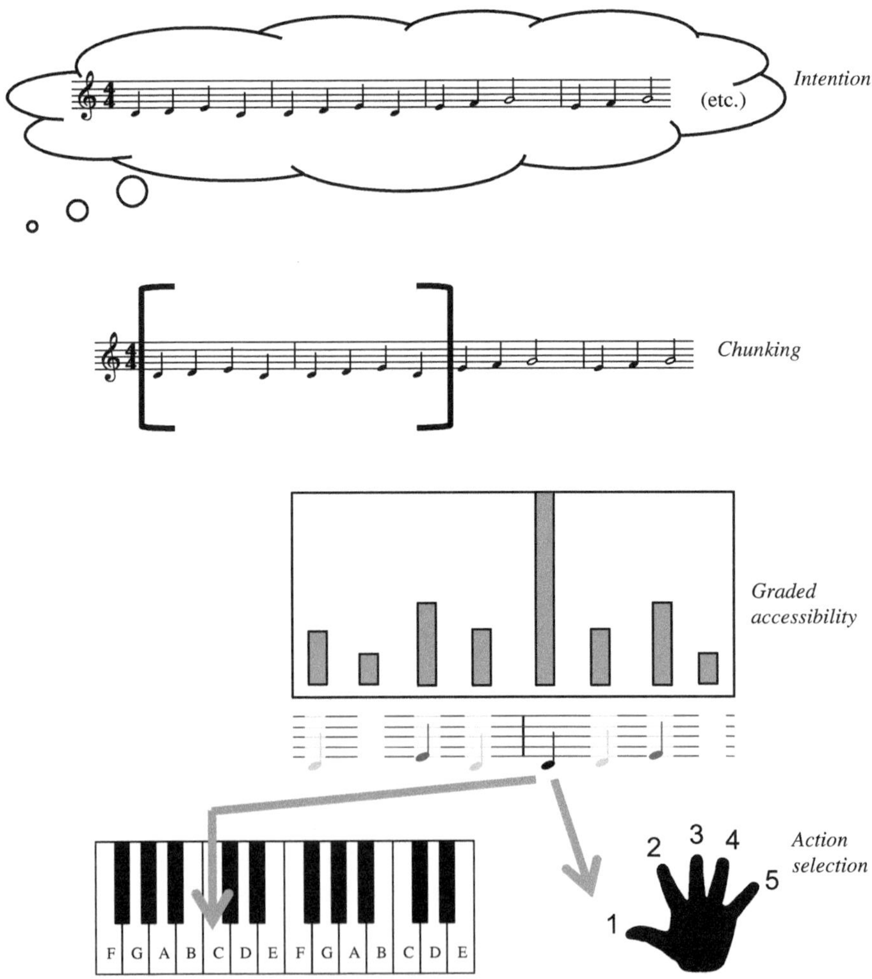

FIGURE 1.1 The formation of an action plan in music performance.

notation for part of an intended melody (here, "Frère Jacques") in a "thought bubble," to convey the fact that the notation is being used to represent an imagined auditory pattern. In this book, when I refer to the *planned melody* in a performance, what I refer to is a produced melody that realizes this intention.

An important consideration regarding the role of imagery is that auditory images, like the music we perceive, are not accessible all at once. Auditory images are scanned in real-time, similar to the way visual images may be scanned (Halpern, 1988). Thus intentions may involve fully formed auditory images stored in long-term memory that are scanned through the next step in the process of performance: the formation and updating of an *action plan*.

Action plans act as an intermediate step between intentions and action, and also an intermediary between perception and action. First, action plans involve *incremental retrieval* from the intended sequence. During incremental retrieval an action plan is continuously updated while execution occurs. Incrementality is important for the production of any complex sequence, including speech and music (Wheeldon, Meyer, & Smith, 2002). Consider reading out loud from a book. The book provides a concrete abstract representation of your intention, which will be spoken prose that your audience can hear. Although you will execute your speech one word at a time, if you are a good reader your eyes will not fixate on the present word but instead will scan several words ahead. This allows you to form an action plan based on future words while you are executing the current word. This allows a reader to avoid common pitfalls like inappropriate pausing at the end of a line, failing to parse sentences appropriately, etc. A good reader plans her reading incrementally. In piano performances, the effects of incrementality can be seen in finger movements, which are initiated some time before the execution of the key press (Dalla Bella & Palmer, 2011; Engel, Flanders, & Soechting, 1997; Loehr & Palmer, 2007).

How does incremental retrieval work? Two key mechanisms have emerged in the literature: *chunking* and *graded accessibility*. Chunking is a well-known mechanism for coding information, proposed by George Miller as a way of enumerating the number of information channels available for short-term memory (Miller, 1956). Chunking, or at least an analogous process of segmentation into smaller units, also plays an important role in action planning for music performance. Patterns of timing variability (Van Vugt, Jabusch, & Altenmüller, 2012), retrieval failures (Palmer & van de Sande, 1995), and practice strategies (Williamon & Valentine, 2002) all suggest that musicians organize music into segments based on the hierarchical structure of music, most likely based on phrasing (cf. Lerdahl & Jackendoff, 1983). For instance, when pianists make serial ordering errors in music, the misplacement of musical *events* (notes or chords) tends to occur within a musical phrase, only rarely crossing the boundary into an adjacent phrase (Palmer & van de Sande, 1995).[1] In Fig. 1.1, the process of chunking is represented by placing brackets around a plausible "chunk" within the music being represented.

One difficulty in applying traditional chunking theories to music is that chunks in standard memory models involve all-or-none retrieval. Remember that the idea of chunking was originally introduced to explain why we have no more problem retrieving a list of words with many letters (apartment, refrigerator, automobile), than one with words having few letters (car, house, stove). This is because information is coded as discrete chunks, so what you recall is a word and not letters within a word. Retrieval of a musical phrase is not directly comparable to this scenario. Thus, for music there is *graded accessibility*. One model of graded accessibility, which plays a central role in this book, is the *range model* of planning (Palmer & Pfordresher, 2003; Pfordresher, Palmer, & Jungers, 2007). According to the range model, graded accessibility of events is based on two factors. First, one has more access to surrounding events based on how far away they are, or their *serial distance*. This factor is further influenced by tempo; at slower tempos performers can integrate a broader range of events into their plan. The second factor is *metrical similarity*: during a performance one has more access to events that are associated with metrical accents that are similar to the present position. Thus, if you are currently on a weak beat, you will have more access to events at other weak beats. Fig. 1.1 displays a typical pattern of activations that reflects both of these components as a bar graph. Below this bar graph is a redrawing of the notated "chunk" that reflects the relative accessibility of each event based on its vividness in the notation.

According to the range model, retrieval of events in a sequence relies on sequential associations. Thus, some access to the surrounding context is important for correct retrieval to occur. Thus, the accuracy of retrieval is based on having access not just to the current event but also to the surrounding events (Pfordresher et al., 2007). Associations from the current event to surrounding events help to consolidate the sense of one's current position, which facilitates correct retrieval. When serial ordering errors do occur, patterns of serial ordering errors can be predicted based on how active or accessible different events are.

[1]For our purposes, an event is a single rhythmically defined unit in a sequence: a note, an isolated chord, a handclap, etc. Note that events need not be based on a single performer—a synchronized orchestral blast is an event—although most of this book will consider solo performances until we get to Chapter 9.

Finally, the action plan must lead to action execution. In music performance, action execution occurs across multiple dimensions. Fig. 1.1 represents the process of execution in piano performance. Here, two critical dimensions comprise the selection of the appropriate hand and finger (effector selection), as well as a spatial target on the keyboard. These dimensions are entirely independent of each other—one can press any key with any finger of either hand. The process of matching a pattern of spatial transitions on the keyboard to a pattern of fingering involves complex ergonomic considerations (Parncutt, Sloboda, Clarke, Raekallio, & Desain, 1997; Sloboda, Clarke, Parncutt, & Raekallio, 1998), although the experiments summarized here typically simplify this consideration for the purpose of focusing on perception/action associations.

In this book, I focus on the implications of this planning process for the intersection between perception and action. Note that the process of planning described here integrated perception and action. Although there are undoubtedly different specialized neurocognitive mechanisms devoted to the fine details of auditory perception and motor control, the core of music performance involves a *common code* that combines sensorimotor information (Hommel, 2009, 2015; Hommel, Müsseler, Aschersleben, & Prinz, 2001; Prinz, 1997). The use of this integration is what binds performer and audience together, and gives music its sense of physicality (cf. Leman & Maes, 2014; Maes, Leman, Palmer, & Wanderley, 2014). As we will discuss, this integrative nature can also cause difficulty for musicians in certain situations.

TIMING IN MUSIC PERFORMANCE

Timing is a critical feature of music. In fact, as I discuss further in Chapter 7, the presence of temporal repetition may be a defining feature that distinguishes music from other auditory experiences, including speech. Of course, timing was not absent from my discussion of intention, planning, and execution. The graded access illustrated in Fig. 1.1 stems from the presence of a metrical accent pattern, as well as a dominant rate of performance (i.e., tempo). But we have not yet considered *how* timing is established and maintained during a performance.

Following Pressing (1999), I will here describe a mechanism for timing that integrates two dominant approaches in the literature: a perspective that stems from Dynamical Systems theory based on the proposal that *endogenous oscillators* provide a basis for timekeeping, and a perspective that stems from *stochastic information-processing* models.

I adopt the idea of an endogenous oscillator to account for the deterministic process of maintaining a regular underlying "beat," which is critical for understanding the role of auditory feedback in music performance. The core of timekeeping here is a recurring oscillation that establishes the preferred rate at which an individual will engage in rhythmic activity, as well as the rate at which a musical sequence will feel intermediate, not too slow or too fast (Drake, Jones, & Baruch, 2000; McAuley, Jones, Holub, Johnston, & Miller, 2006; cf. Fraisse, 1982). The task of keeping time in a performance requires that one adapt the rate of this internal oscillator as needed. Skilled musicians become adept at doing this, although in all musicians there may be a tendency to drift in time toward one's most comfortable tempo, leading to speeding up or slowing down as one performs. At the top of Fig. 1.2, the dark circle represents an oscillating *limit cycle*: a stable state based on a recurring period. The curved paths that move toward this circle indicate how the cycle draws behavior from any starting point into its path. For a real-world example, think of the pendulum on an old grandfather clock. The swinging motion of the pendulum will take on the same recurring period regardless of where you start the motion.

Below this circular representation I have plotted a sinusoid, which results from stretching out this cycle over time. Here we can see how the oscillating limit cycle may be related to physical movement, such as the rising and falling of a finger if you play a repeating quarter note on a piano key. The oscillation itself is defined by a period, or rate of recurrence, and phasing, which determines the start points of the oscillation. The start point of oscillations may be associated with intended contact

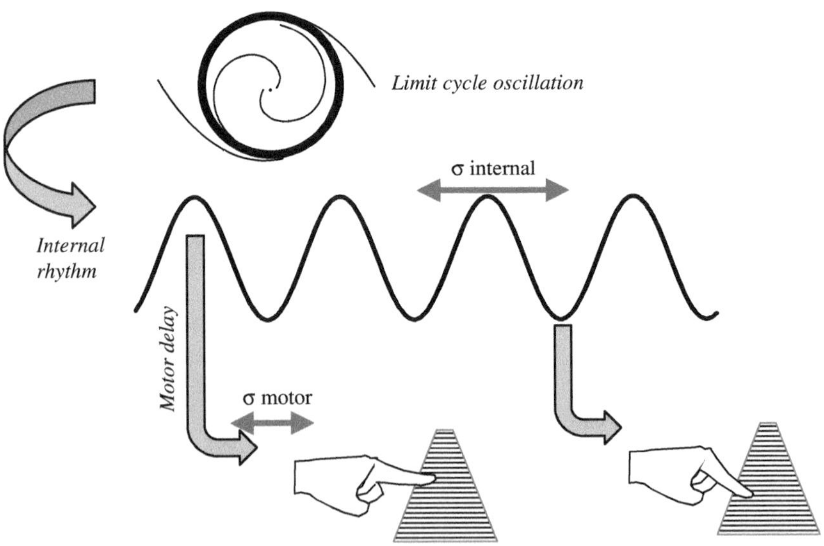

FIGURE 1.2 Temporal control and variability sources in music performance.

times for finger taps on a surface (as in Fig. 1.2), or syllable onsets in singing, or even points in time at which one expects to hear the next musical event in a melody (Large & Jones, 1999). The fact that this internal rhythm may be used for both perception and production of timing is a critical point, which I will describe in more detail when we consider timing relationship between perception and action in Chapter 3.

Timing in music also exhibits variability that seems to have a random basis. Because of that, Fig. 1.2 also suggests that some random (stochastic) variability may influence the timing of the internal oscillation, symbolized by Greek sigma (σ). In addition, based on evidence from studies of variability in rhythmic tapping, the framework I propose includes a second source of variability, based on the delay in time from the brain's command to produce a movement to the time at which the actual muscle contractions occur (Wing & Kristofferson, 1973a, 1973b). Quite possibly the ability to register the timing of an auditory sequence in perception may also be influenced by random noise.

Of course, timing in music involves more than just maintaining the beat. Performers also vary timing from a regular beat to convey expressive intentions. For instance, performers often slow down slightly at the end of a phrase, and speed up toward the middle of a phrase (Gabrielsson, 1999; Palmer, 1997). Although expressive timing is highly important to the aesthetic experience of music, it will have relatively lower importance in the present volume, with the exception of Chapter 9, in which we consider how duets and ensembles synchronize expressive timing.

PERCEPTUAL AND AUDITORY FEEDBACK

Another similarity across music and speech has to do with the fact that in both domains the actions themselves are somewhat inconsequential. Communication for the most part comes from the *auditory* consequences of actions. Consider a situation in which you are playing a rather old piano that's in a condition of disrepair. One of the keys that you need to use is horribly mistuned, and distracting to hear. Does an accomplished pianist hold fast to the action sequence, and continue to use this key and cause disruption of the sound pattern? Certainly not. A skilled performer would do anything possible to avoid using this key, which could be omitting the notes or even replacing that pitch with another one that fits the overall melodic and harmonic context. The audience came to hear sounds, primarily. Of course musicians' actions contribute to musical experience, otherwise we would not take the trouble to combine auditory and visual experiences at a concert

(Livingstone, Thompson, & Russo, 2009; Thompson, Russo, & Livingstone, 2010; Vines, Krumhansl, Wanderley, Dalca, & Levitin, 2011; Vines, Krumhansl, Wanderley, & Levitin, 2006). Nevertheless, these movements typically have an indirect relationship to actions used to produce notes (Wanderley, 2002) and may reflect larger timescales of musical organization than the actions that produce individual notes and chords. For instance, listeners respond strongly to oscillations of the trunk exhibited in expressive piano performance (e.g., Davidson, 1993), whereas finger movements are often impossible to see. More to the point, you don't find people watching performances with the sound turned off, whereas it is very common for people to listen to music without accompanying video.

The auditory consequences of actions are therefore critical to performance; these consequences are here referred to as *auditory feedback*. Thus, in contrast to the *planned melody* that a performer intends to produce, I also refer to the *feedback melody* as the melody that accompanies the performance, which may or may not be interpreted as directly resulting from actions that make up the performance. Of course there are many forms of *perceptual feedback*, with auditory representing just one form—albeit perhaps the most important form of feedback for music. Perceptual feedback encompasses all modalities, and in most circumstances we receive rich multimodal feedback. Thus, in addition to auditory feedback a pianist will receive tactile feedback (touching the key), kinesthetic feedback (movements of the muscles), visual feedback, and possibly even vestibular feedback based on head movements. All these sources of feedback are of interest. However, I focus on auditory feedback because it is most likely to be associated with the goals of movements, or the performer's intentions, and because of the domination of auditory feedback in the existing literature.

Oddly enough, a major theme of this book concerns whether the term *auditory feedback* accurately reflects the role of sound in music performance. I use this term to avoid confusion, given its longstanding use, although I agree with Peter Howell (e.g., Howell, 2004a) that auditory feedback may not be accurate in a technical sense.

The initial adoption of the term *feedback* probably seemed like an appropriate descriptive term given the kind of manipulations that were used initially. Delayed auditory feedback (DAF), which will be discussed in detail in Chapter 3, was introduced in the 1950s by researchers interested in sensory control over speech. As the name suggests, DAF involved having participants speak while the sound of their own voice was delayed by small amounts—usually fractions of a second. This effect is much like the aforementioned example of speaking into a telephone and hearing a delayed copy of your voice. Such manipulations hardly beg the question of whether the manipulation is of feedback, because the heard sounds come directly from one's own voice.

In subsequent years, however, the importance of "feedback" in this manipulation has become less clear because similar effects to DAF can arise when sequence production is accompanied by any concurrent sound that conflicts with the rhythmic structure of produced speech. Such situations are not hard to imagine from daily life either. Consider the experience of trying to sing in a choir next to an individual whose part contrasts sharply with yours in its rhythmic and pitch properties. Many individuals find the auditory information from their neighbor disruptive even though it is not "feedback" per se. Thus, my use of the term *auditory feedback* is partly due to maintaining continuity with past literature, more so than a commitment to the theoretical implications of the term.

THE ROLE OF FEEDBACK IN MODELS OF PERFORMANCE

Much of this book will be given over to consideration of models for the relationship between perception and action, and whether these models can account for the role of sound in music performance. It will help, however, to discuss in broad terms a few canonical approaches that help to frame the discussion that follows.

Closed-Loop/Cybernetic Approaches

In the 1950s, when studies of DAF began, the leading theoretical paradigm in motor control came from cybernetics (Brown & Campbell, 1948; Wiener, 1948, 1961). The influence of this paradigm can be seen in the discussion sections of the first papers, in which DAF effects were reflexively interpreted in light of *feedback control*. It is important, therefore, to consider what feedback control means. Fig. 1.3 illustrates the distinction between a feedback control account, also referred to as *closed-loop* control based on an engineering metaphor, and an *open-loop* account. In these models there is a separation between planning and execution. Whereas planning involves retrieving a goal response from memory, execution involves the physical implementation of that goal. Execution also involves the control of an external device like a musical instrument. As you can see, the only difference in these models has to do with how output from the system is used. In an open-loop system, the output does not influence system control in any way. Importantly, according to such a model, alterations to the output (like DAF) should not influence either planning or execution.

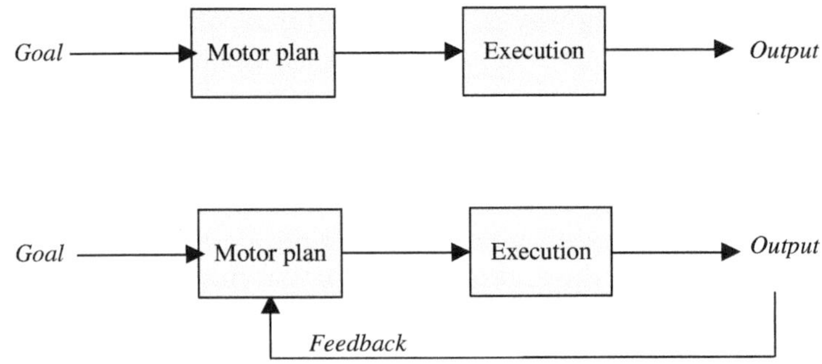

FIGURE 1.3 Highly simplified control system reflecting open-loop (top) versus closed-loop (bottom) control.

By contrast, in a closed-loop feedback control system, the motor plan is influenced directly by the relationship between the intended outcome and feedback from the executed action. For instance, say you are reaching out your arm to press a button, and for some reason your arm goes a little off course (e.g., you have just finished a round of 100 bicep curls at the gym). In a closed-loop system, the discrepancy between the intended location of your finger and the goal state would register as an error, and some corrective activity within the motor plan would occur.

Initially the effect of DAF seemed to confirm the predictions of a closed-loop cybernetic system. By this logic, disruption occurs because the delayed signal is interpreted as an error by the system, and in response the system initiates corrective changes to the motor plan in an unsuccessful attempt to reduce this error (unsuccessful because, of course, the delay can never be overcome). For instance, one of the strongest effects of DAF is that speech slows when it is introduced. This response may occur because a speaker attempts to slow down speech commensurately with the apparent slowing of speech onsets.

In the 1950s and 1960s, several closed-loop theories were proposed to account for effects of DAF on speech (e.g., Chase, 1965; Fairbanks, 1954; Lee, 1950). All of these involve elaborations on the basic architecture shown in Fig. 1.3. For instance, Fairbanks (1954) proposes that effector control of speech leads to a set of feedback channels, including auditory information (air and bone conducted) and body senses (tactile and kinematic), all of which are aggregated and compared with a dynamically changing goal state (intended speech output). Any discrepancy between the goal state and the aggregated feedback channels can lead to an error signal, resulting in a command for the effector control system to alter its behavior. An interesting component of this model is the idea that different feedback sources may conflict with one another, and that this conflict may play a role in disruption. Later we will see this idea developed further in an open-loop model proposed by Peter Howell.

Karl Smith, who published highly influential studies of delayed visual feedback, proposed another kind of closed-loop model that operates on higher-order sequential principles, referred to as the *Neurogeometric Theory* (Smith, 1962). According to this view, delayed feedback disrupts sensory change-detectors, referred to as *internuncial neurons*. This proposal situates that the cause of disruption at sensorimotor incongruence in transitions between motor versus sensory states causes disruption, rather than absolute differences. As applied to speech and visual tracking—the contexts Smith bore in mind—transitions include all the time-varying postural shifts that occur both within and across events. In the context of piano performance, the Neurogeometric Theory would take into account movements between event onsets and not just the timing of individual onsets. Delayed feedback disrupts piano performance not because there is a discrepancy of onset timing, but because the tone onset interferes with neurons being used to plan the time-varying movement trajectory. As we will see, Smith's view succeeds in accounting for some but not all altered feedback effects.

The closed-loop view predicts a strong, critical role for auditory feedback in music performance. In a particularly extreme version of this view, discussed further in Chapter 2, recall relies entirely on the presence of auditory feedback. As is often the case for an initial theory, the closed-loop perspective has many deficiencies in accounting for the role of auditory feedback. I will discuss its shortcomings in considerable detail through the next several chapters, but for now we will move on to discuss two other models that help frame later discussion.

Opening the Loop: The Displaced Rhythm Hypothesis and EXPLAN

Based in part on the limitations of the closed-loop view, Peter Howell and colleagues proposed that auditory feedback effects are limited to temporal synchrony (Howell, Powell, & Khan, 1983). As we will review further in Chapter 3, this view was inspired primarily by altered feedback effects in speech, although it has important implications also for music performance. This view is in some respects like the simple open-loop model shown in Fig. 1.3, in that it proposes that the planning of sequences is not influenced by feedback. Instead, Howell proposes that feedback influences a separate timekeeping mechanism that guides the initiation of execution.

In Chapter 3, I will discuss the *Displaced Rhythm Hypothesis*, which Howell introduced as a way to explain the role of auditory feedback and effects of DAF. Howell further developed this perspective in the

EXPLAN (EXecution and PLANning) model of speech production (Howell, 2001, 2002, 2004a, 2004b; Howell & Au-Yeung, 2002). This model was introduced primarily to account for stuttering. However, because stuttering may be associated with feedback processing (further discussed in Chapter 7), EXPLAN incorporated an account of feedback processing, which is illustrated in Fig. 1.4. The central claim of this model is that planning and execution occur in parallel, following the aforementioned principle of incrementality. However, the time at which each event in a sequence (e.g., each word) is planned occurs independently of the timing for each event's production. An independent timekeeper that regulates the rate at which different events are executed guides the timing of speech production.

Fluent production occurs when the timing aligns such that the execution of a word occurs right after the plan for that word has been fully formed, as in the top half of Fig. 1.4. Problems arise if a plan for a produced event has not been fully formed at the time when execution

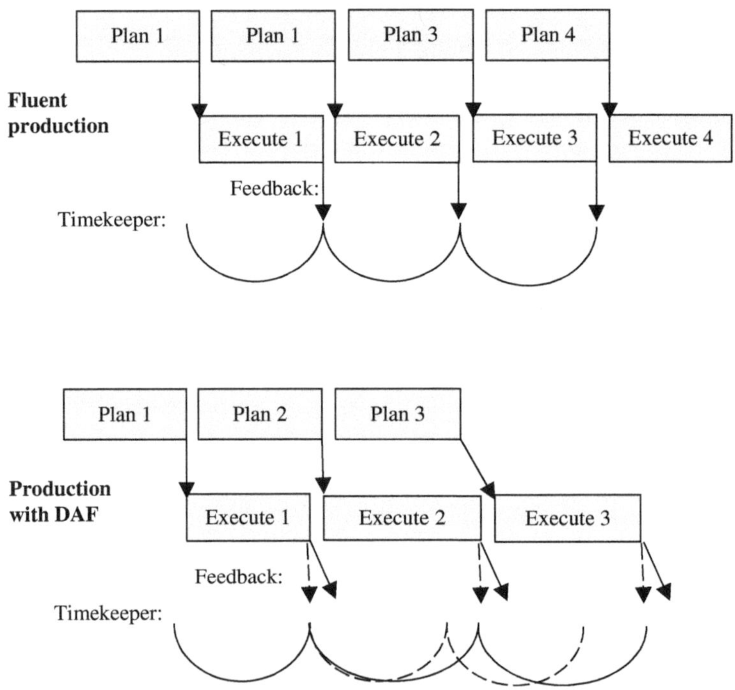

FIGURE 1.4 The EXPLAN theory of speech planning and role of auditory feedback. Numbers indicate sequenced events (e.g., syllables). Dashed arrows indicate anticipated timing of events given the timing of execution, whereas dark arrows indicate the timing of auditory feedback. The effect of DAF on the timekeeper is shown by comparing the intended timing of execution (dashed line) to the actual timing of execution that results from asynchronies between auditory feedback and the intended timing of actions.

begins. Assume you want to say, "the boy went to the store." If the plan for the word "boy" is not fully formed by the time you start to produce this word, stuttering can arise. This is common for pairs like "the boy," given that function words are executed rapidly. As a result, by the time one finishes saying "the," and starts to execute the next word ("boy"), it is possible that the plan for this content word—representing a larger class of words to choose from—may be unavailable. Thus stuttering, according to this model, occurs when the timing for plans is paced at a rate that is slow relative to the rate at which execution is attempted. That's why even fluent speakers tend to stutter when they are nervous and try to speak too quickly.

All this leads to the EXPLAN account for effects of altered auditory feedback. According to the model, auditory feedback does not interact with planning, only execution. Furthermore, disruptive alterations affect an external timekeeper that is responsible for regulating speech rate. The effect of disruption is to slow this rate, which is probably the most common effect of DAF as we will discuss more in Chapter 3. Howell (2004a, 2004b) suggests that the timekeeping mechanism in EXPLAN is based in the cerebellum (see Fig. 3.7).

A critical point about Howell's prediction is its stark contrast with the closed-loop view. Another noteworthy feature is its rhythmic basis, which for music seems to be a critical factor in perception/action associations. Although EXPLAN was developed primarily to explain the intersection of perception and action for speech, Peter Howell has a longstanding interest in music cognition (cf. Howell, West, & Cross, 1991; West, Cross, & Howell, 1985) and an extension of this framework is straightforward in principle (but see Howell, 2004a and Chapter 7 for further discussion). Despite its merits, as will see in Chapter 4, EXPLAN is too limited to account for all aspects of perception/action associations in music.

A Shared Network: Node Structure Theory

During the 1980s, a new class of cognitive models arose that integrated the possibilities arising from computer modeling with longstanding interest in how neural networks in the brain build associations. These models are referred to as *connectionist*, or *neural network* models. During this period, Donald MacKay proposed a comprehensive theory of perception and action based on a shared connectionist network of hierarchically organized associations. The result, *Node Structure Theory*, was designed to account for a variety of phenomena from speech perception and production, although the framework is designed to generalize potentially to other sequential behaviors. Node Structure Theory has almost all the elements of a workable theory for the role of feedback in the production of complex sequence. It fits into a broader framework of models that propose shared perception/action representations (e.g., Hommel et al., 2001), but is uniquely well suited for perception and action in music given its focus on the production and perception of complex auditory sequences.

According to this theory, perception and action both rely on a hierarchical network comprising interconnected nodes. There are three kinds of nodes, which form three interacting systems. The most well-known system comprises *content nodes*. Content nodes convey domain-specific meanings at different hierarchical levels. For instance, a linguistic content node may convey an intended proposition (moldy jam tastes bad), a word (jam), or a phoneme (/j/). During production, activation starts at a high hierarchical level. Activation at this high level "primes" many candidate nodes at lower levels, and from these candidates the most highly primed node is then activated, which in turn primes nodes at still lower levels. During perception, activation spreads in the opposite direction. For music, content would likely involve the hierarchical arrangement of nodes and chords into phrases, as discussed in music theory (e.g., Lerdahl & Jackendoff, 1983). Fig. 1.5 illustrates this kind of arrangement for the first three phrases of "Frère Jacques" ("Are you sleeping").

The two other systems of nodes are more central to the research I summarize in this book. They govern how content is constructed in production or comprehended in perception, and reflect the importance of how content is organized in time. One set of *sequencing nodes* determines how content must be ordered according to rule systems of language. For instance, these nodes ensure that a planned adjective precedes a planned noun ("happy puppy") when one plans to produce the associated noun phrase. Within music, of course, this system must follow different constraints, yet in both music and language content must be treated separately from sequencing. In music, a given pitch event may be used many times in a melody, and yet will assume different roles. For instance, the first pitch of "Frère Jacques" (C, in Fig. 1.5) is repeated in the fourth position. Fig. 1.5 presents a serial order based on laterally inhibitory connections, as is common in speech production models (Dell, Burger, & Svec, 1997). Activation of the first sequential position inhibits all remaining positions. When it comes time to produce an event, the most likely event is thus the content associated with position 1. Once that content has been activated its activation diminishes, leading the second sequential node to be activated, etc.

One feature of Node Structure Theory that figures strongly in the present volume is its separation of sequencing and timing. The importance of this distinction is discussed further in Chapter 5. Timing nodes

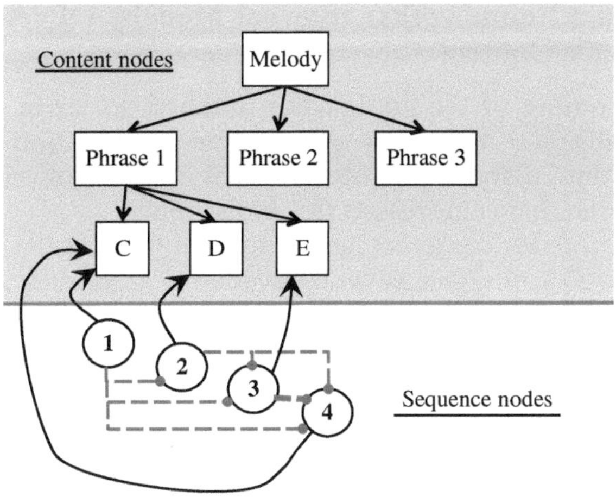

FIGURE 1.5 Illustration of Node Structure Theory as applied to a musical context: the first phrase of "Frère Jacques" (cf. Fig. 3.2 from MacKay, 1987, p. 54). Content nodes are shown as *rectangles* over the *gray background*, whereas sequence nodes are shown as *circles*. A subset of the phrase hierarchy within the entire melody is shown at the top two levels of content nodes, as well as the pitch content within phrase 1, denoted by standard pitch labels. The four sequence nodes have lateral inhibitory connections (*dotted gray lines*) that lead to ordering of the associated content nodes, with sequence/content associations illustrated by *curved arrows*. Timing nodes (not shown) would determine the activation timing of successive sequence nodes.

dictate the rate at which sequences are produced by designating the time at which sequential nodes are activated. The theory assumes an indefinite number of timing nodes, with different timing nodes being used for different timescales within a sequence (faster timing at the phonological than the lexical level) or for events associated with different durations (e.g., content vs function words). Although Node Structure Theory does not specify a mechanism for timekeeping, one analogue would be a bank of oscillators each tuned to a different rate (as opposed to the single rate shown in Fig. 1.2). The details of this kind of timing mechanism will be discussed in Chapter 3. Fig. 1.5 does not show any timing nodes, but for this phrase the theory would claim that a single timing node suffices to activate each sequential node at regular time intervals given that the rhythm is *isochronous* (each duration is equivalent).

Thus, Node Structure Theory provides a wide-ranging framework for understanding many critical issues that arise in this book. In Chapter 3 I will detail the specific predictions of Node Structure Theory as it applies to the role of auditory feedback. I will argue that a broadened version of this kind of model could provide a good account for known phenomena.

Computational Frameworks: Internal Models, DIVA, and State Feedback Control

One shortcoming of the approaches outlined so far is that they are framed verbally and thus lack quantitative specification (see Finney, 1999, for further discussion). More recent models of motor control, including speech, have addressed this limitation.

The *Internal Models* construct has been highly influential for the study of motor control in the past several decades (e.g., Jordan & Rumelhart, 1992; Kawato, 1999; Kawato, Furukawa, & Suzuki, 1987; Wolpert, 1997; Wolpert, Ghahramani, & Jordan, 1995). In a basic sense, Internal Models reflect the basic claims of motor schema theory (Schmidt, 1975), in that the performer internalizes some kind of rule system that can be used to relate planned actions to the perceptual consequences of those actions. One difference is that recent incarnations of Internal Models include Bayesian estimation as a computational framework to update the model's predictive behavior (e.g., Körding & Wolpert, 2004, 2006). Another difference is that Internal Models separate two routes for perception/action translation. Whereas *Inverse Internal Models* work backward from the perceptual representation of an intended goal to form an action plan, *Forward Internal Models* translate the current action plan into the most likely predicted output. The output of a Forward Model can be compared to perceptual feedback in much the same way as in closed-loop approaches like Fairbanks (1954).

A related computational model for learning of sensorimotor association in speech is the DIVA model (for "Direction of Velocities into Articulators," Guenther, 1994, 1995, 2006). This model comprises a highly detailed sensorimotor map of the speech system based on both computational and neuroanatomical principles. As such it can plausibly be applied to singing, although its application to other forms of music performance is less clear. According to DIVA, speakers develop an Internal Model that maps articulations to auditory feedback through simple associations (in a "babbling" phase of learning) and through error-based learning. Like closed-loop approaches, DIVA uses auditory feedback as an error signal to adjust motor planning for speech.

Recent models have capitalized on similarities between the Internal Models construct and DIVA to develop integrated frameworks (Hickok, Houde, & Rong, 2011; Houde & Nagarajan, 2011; Max, Guenther, Gracco, Ghosh, & Wallace, 2004). The most applicable model for our present purposes is the *State Feedback Control* model of Houde and Nagarajan (2011). This model integrates many critical details of DIVA into an Internal Model framework in which error correction is based on discrepancies between auditory feedback and the expected outcome from the Internal Model. What distinguishes this from other Internal

Model approaches is the level of computational detail both with respect to processing components and mechanisms of error correction. This assumption is also critical to the hypothesis I propose next.

Although these models share many assets they also suffer from a common limitation, in that none of them provide a framework for higher-order sequential organization. As we will see, this kind of organization determines the role of sound in music performance to a large degree. Although it is possible to string together sequences of Internal Models in a chain (Wolpert & Kawato, 1998), some kind of higher-order structure is necessary to account for the organization of serial order (Lashley, 1951). Insofar as serial order plays a role in musical perception/action associations, therefore, one must move beyond these computational approaches.

In contrast to other models, the neural underpinnings of these computational approaches have been well developed. The DIVA model has been associated with connections between the primary motor cortex and higher-order auditory areas in the brain (Guenther & Vladusich, 2012), whereas Internal Models—similar to the timing component of EXPLAN—have been linked to areas within the cerebellum (Wolpert, Miall, & Kawato, 1998).

The Present Approach: Hierarchical Perception/Action Coordination

In this book I will develop an approach that synthesizes elements from existing theories of perception and action, along with our current understanding of the mechanisms that contribute to music performance discussed earlier. My proposal is entitled the *Hierarchical Perception/Action Coordination* hypothesis, or H-PAC. I will develop a preliminary model based on this hypothesis through the course of this book; here I reflect on its general features and how they relate to existing models.

The centrality of coordination arises in part from Howell's EXPLAN model, although coordination in H-PAC stems also from the metrical organization arising from pitch sequencing. The hierarchical aspect of H-PAC comes from the separation of sequencing and timing, inspired by Node Structure Theory, although the specific mechanisms of H-PAC differ from those proposed in Node Structure Theory. Like EXPLAN, H-PAC considers effects of altered feedback to be based on temporal coordination, although unlike EXPLAN it proposes that disruption of planning as well as execution can be influenced by these rhythmic interactions. Because sequencing extends across a broad timescale in comparison with the model's timing mechanism (see Chapters 3 and 4), they form a two-level hierarchy. In tandem, these components lead to the retrieval of sequence events at the appropriate time in production, and the perceiver's ability to track and comprehend unfolding events in perception (see Chapter 5).

At a lower hierarchical level, I hypothesize that a mechanism for sensorimotor translation provides a direct interface between perception and action at the level of individual events. This component of the model functions similarly to closed-loop and computational approaches to perception and action. The main role of this component comes about in fine-grained control in production, including pitch control in musical tasks like singing, and in the motor coding of perceptual input. Another way in which H-PAC resembles these earlier approaches is in its claim that auditory feedback interacts with a representation of action planning, as opposed to influencing execution directly, as in EXPLAN.

PARADIGMS AND MEASURES

Most of the studies discussed in this book share methodological features. I detail some of the key paradigms and measures here so that later discussions can focus on important conceptual issues, or methodological advances that are of particular interest. Readers who are familiar with these kinds of methodological details can skip this section.

By far the most prevalent paradigm in this book involves the use of *altered auditory feedback* mentioned earlier. The participant here produces a melody (or any kind of auditory sequence) while experiencing either normal (unaltered) or altered auditory feedback. The type of alteration can be of any sort, although primarily alterations have influenced the synchronization of feedback with actions (Chapter 3) or the pitch content of feedback (Chapter 4). At issue is whether a given alteration disrupts production, along with the extent and type of disruption that results. Importantly, not all alterations lead to disruption. For instance, removal of auditory feedback does not tend to disrupt production although it can lead to other negative effects (Chapter 2).

What constitutes *disruption*? Most obvious, disruption can emerge as an increase in produced errors. Assuming that the performer intends to produce an accurate sequence of events, any increase in error frequency due to altered feedback suggests some sort of disruption on the way from intention to action. Occasionally researchers have looked deeper in the types of errors performers make, for instance, addressing what kind of sequence positions are involved in disruption of serial order (Pfordresher & Palmer, 2006), or how often errors involve inserting a new event versus substituting a different event (Pfordresher & Mantell, 2012). But as we will see, not all alterations increase errors. Another complicating factor in using error rates is that errors occur infrequently and reflect a discrete dichotomous variable; both of these features influence how well error rates meet the assumptions of standard statistical tests.

It is also common to measure disruption of timing in performance. One of the first observed effects of DAF was that the rate of speech slowed down, and this remains a reliable effect for many feedback alterations to this date. In music performance the most common measure of rate is the mean of interonset intervals for a performance, removing those intervals that are associated with errors. Performers typically slow down after making an error (Rabbitt, 1966, 1978), and so it is important to exclude these interonset intervals if one wants to analyze performance accuracy separately from timing. The relationship of slowing to "disruption," however, may seem more transparent than it is. The goal of performing music, typically, is not to perform as rapidly as possible but instead to match an intended tempo. Thus, there are really two patterns of performance that may demonstrate disruption, performing too slowly or too rapidly. But the latter effect almost never emerges.

The precision of timing, or timing variability, can also reflect the disruptive effects of altered feedback, although it has been used less often than errors and production rate. Here, one must rule out both errors and production rate as covariates. Timing variability tends to increase as production rate slows (Wing & Kristofferson, 1973a). Thus, to measure timing variability independent of production rate (tempo), researchers typically use the coefficient of variation, which is the ratio of the standard deviation to the mean. This coefficient remains constant if there is a proportional relationship between the two measures. Expressive timing also complicates the interpretation of timing variability. Performers, particularly of classical music, typically vary the tempo as they perform in a deterministic way, to express an emotional interpretation and to highlight phrase boundaries. These expressive nuances increase the variability of timing, but nobody would want to consider such patterns of variability as evidence of "disruption." Researchers typically address this issue by having performers play rather bland pieces of music, with instructions to make timing as precise as possible rather than make the music sound interesting and expressive. That being said, the question of how feedback alterations may influence expressive timing is an inherently interesting one and to date is woefully unstudied. I know of only one exception, Repp (1999), which I will discuss in Chapter 2.

Most existing research, as noted earlier, measures disruption of keyboard performance, which simplifies greatly the measurement of these variables. Because pitch is represented discretely, errors can be identified transparently based on mismatches between the intended event sequence and the performed sequence, using algorithms that account for the minimum number of alterations necessary to match the two sequences (Large, 1993; Palmer & van de Sande, 1993, 1995). Likewise, it is easy to identify onsets in a percussive instrument like the piano. A significant advancement in the 1980s was the development of the Musical Instrument Digital Interface (MIDI) file format, which represents all performance information (e.g., note value, keystroke velocity) as a sequence of discrete numbers. As a result, critical features in the continuously varying pattern of movement (action) and sound can be represented discretely.

More recently, advances in computing have made it easier to measure performance features in singing, but the demands of measuring singing performance remain much higher than in piano. The task of identifying onsets is nontrivial, in part because the intended onset time of sung notes does not consistently associate with a single acoustic marker (Sundberg & Bauer-Huppmann, 2007). The identification of errors presents an even thornier problem. One can label as "error" any pitch that diverges from the intended F0 by a half semitone (50 cents). However, even trained singers often mistune pitches by this much simply based on the challenges of intonation (Pfordresher & Brown, 2017) with listeners usually not exhibiting any awareness (Hutchins, Roquet, & Peretz, 2012). Although it is possible to look at how auditory feedback influences vocal intonation, and I will discuss this literature in Chapter 7, the main focus of this book will be on errors of sequential organization. As such, research concerning how altered feedback disrupts the sequencing of song typically measure the rate at which errors of melodic contour—the direction of pitch change—vary with type of auditory feedback (e.g., Pfordresher & Mantell, 2014; Pruitt & Pfordresher, 2015).

Although a lot can be learned from focusing on discrete features of performance, many labs have begun to study continuous patterns of movement that surround these critical moments through *motion capture*. Tiny "markers" are placed at critical positions on the participant's body; a system of cameras measures the movements of these markers in three dimensions over time. This system supplies a vast amount of sampled data, from which the researcher can analyze the position, velocity, or acceleration of movement within individual markers, or aggregated across markers. Motion capture can focus not only on fine-grained movements of fingers and hands in piano performance (e.g., Dalla Bella & Palmer, 2011; Goebl & Palmer, 2008; Loehr & Palmer, 2007), but has also been used to analyze gross patterns of body movement in dance (e.g., Burger, London, Thompson, & Toiviainen, 2018; Burger, Thompson, Luck, Saarikallio, & Toiviainen, 2013; Toiviainen, Luck, & Thompson, 2010).

The altered feedback paradigm is often critiqued due to its artificiality (e.g., Wing, 1977), and so some defense of it is warranted here. In daily life we take the normal functioning of our nervous system for granted. Yet, the fact that our system typically functions in a given way does not mean that normal functioning depends on every normal contingency in that system. For instance, most people look at their food while they eat. However, it is possible to eat while blindfolded, and doing so may even enhance one's gustatory experience, leading to several restaurants that have people eat in the dark! Aberrations from normal experience and functioning can be informative. They can emerge from unusual cases, as in the study of aphasia and amusia, or from the effects of artificial situations.

Although the altered auditory feedback paradigm will dominate the discussion here, a few other related paradigms will appear from time to time. These other paradigms rely on similar measures of performance, and sometimes similar manipulations, but to different ends. Two paradigms are slight alterations of the standard altered feedback paradigm. One of these involves *sensorimotor learning*, which will figure strongly in Chapters 2 and 6. This paradigm involves the learning of action sequences based on both perceptual and motor factors. In many cases, studies have looked at how nonmusicians form associations between action planning and perceptual feedback through experience. Quite often these studies involve alterations of auditory feedback, but the rationale behind these alterations differs from the standard altered feedback paradigm because the point of these studies is how auditory feedback influences the acquisition of sequence learning, rather than whether altering feedback disrupts ongoing production of an already learned sequence. Another paradigm that often incorporates feedback manipulations, but for a different purpose, is the *joint performance* paradigm, the main focus of Chapter 9, which investigates how performers can synchronize with each other. The joint performance paradigm mirrors the actual practice of performance closely; duet performance is most widely studied although some studies of larger groups exist as well.

A final paradigm addresses perception/action associations in a different way, by examining our ability to engage in *imitation*. Imitation always involves some kind of conversion from perception to action, in an attempt to replicate the kind of sensory output one has perceived. Our ability to imitate thus depends on the same kind of perception/action associations that cause feedback alterations to be disruptive. Of course, imitation differs critically from the altered feedback paradigm because these interactions are desirable and necessary for success. Whereas altered feedback effects reflect perception/action associations that are unavoidable, the imitation paradigm addresses how well we can match fine-grained details of our perceptual experience to an action plan. Another important difference from the altered feedback paradigm has to do with timing. Whereas auditory feedback studies focus on concurrent perception/action associations, these associations are separated in time during imitation. Recently the imitation paradigm has expanded within music cognition due to an increased interest in singing accuracy, and particularly in the problem of poor-pitch singing. Although the main focus of this book is on auditory feedback, I will from time to time consider associated research that uses the imitation paradigm.

Cognitive Neuroscience and Neuroimaging

To this point, I have only considered behavioral paradigms, which will constitute the bulk of the research discussed in this book. However, research addressing the neural representation of perception and action has increased dramatically in recent years and I will also include results from these studies. Broadly speaking, the neuroscience research falls into two categories, based on measures that are best suited to address different kinds of questions about the brain.

One category includes studies that focus on *localization of function*, or the areas in the brain that play some role in how perception and action interact. The most prevalent technique used to address this question is *functional magnetic resonance imaging*, or fMRI. fMRI measures changes in the oxygenation of blood flow that occurs following the coordinated activity of neurons in a certain area of the brain. Images relating to these hemodynamic changes are superimposed on a high-resolution anatomical MRI image that gives a clear picture of where some kind of activity may occur. An older measure, *positron emission tomography* (PET), uses a radioactive trace in the blood to measure changes in blood flow that follow brain activity. Both of these measures offer high spatial resolution, thus providing good measures of localization, but are not well suited to measure the timing of brain activity. Classic studies using these measures rely on the *subtraction method*, in which brain activity during some kind of baseline control condition (e.g., performing a melody with normal auditory feedback) is subtracted from activity during an experimental treatment condition (performing with altered auditory feedback). Areas in which this difference is statistically significant are thought to reflect the influence of the experimental manipulation. More complex methods of analysis have emerged in recent years, such as the use of patterns of brain activity as way to classify the relationship between brain states and different cognitive tasks, and the use of correlations across brain areas to measure connections within neural networks.

The second methodological category focuses on *neural dynamics*, or the time course of neural activity, and use measures that prioritize temporal resolution of measurement as opposed to the spatial precision boasted by fMRI and PET. The most common technique is the use of *electroencephalography*, or EEG, which involves the use of sensors on the skull to record fluctuations in electrical potentials associated with the coordinated flow of ions in the extracellular fluid surrounding neurons. When neurons fire, the charge in extracellular fluid alternates between positive and negative polarity, and when populations of neurons fire together these changes summate to levels that are measureable via EEG. The most common measure of EEG in literature to date is the *event-related potential* (ERP), which is the average change in EEG that is associated with the onset of a critical event. For instance, in Chapter 4 we will discuss how ERPs respond to alterations to the pitch content of auditory feedback during piano performance. A downside of EEGs and ERPs is that the spatial location of the activity is hard to determine, whereas the timing of measures is closely related to neural activity (within a millisecond). A related, but much more expensive, technique is *magnetoencephalography* (MEG), which uses the magnetic field associated with changes to polarity. MEG yields better spatial resolution than EEG/ERPs (thought still not as good as fMRI), while also boasting high temporal resolution.

CONCLUDING REMARKS

This chapter provided a general introduction to the theoretical questions that permeate this book, the methods used to address these questions, and the theoretical models that have been used to explain the relationship between action and sound to date. All of this helps to set the stage for more detailed chapters to follow. In the next chapter, I address an important basic question: whether auditory feedback is actually necessary for fluent music performance to occur.

CHAPTER

2

Do We Need Auditory Feedback? If Not, Why Not?

The title of this chapter poses a question. The glib version of my answer would something along the lines of, "not as much as you might expect." The presence of auditory feedback can significantly facilitate learning of a melody, although its presence does not seem to be necessary for this learning to occur. Furthermore, once learning has finished, auditory feedback plays a surprisingly minor role. Masking or removing it can affect performance, particularly when pitch control is demanding as is the case in singing, but its removal in no way disrupts a performer like other manipulations of feedback that I will discuss in subsequent chapters.

As noted at the start of Chapter 1, perhaps the most unappreciated musical feat exhibited by the Beatles was their ability to perform before large arena crowds during the heyday of "Beatlemania." At the time, the audio equipment used by the band did not incorporate the powerful monitor speakers that performers use today in order to hear themselves. As a result, the group could barely hear themselves over the deafening roar of the crowd. The Beatles were famous for their intricate three-part harmonies (This Boy) and counterpoint (She Loves You). One could imagine that near absence of auditory feedback—the result of the acoustic setting in which they performed—may have ruined their vocal performances. Yet, when you listen to recordings like "Live at the Hollywood Bowl" the performances hold up surprisingly well. The Fab Four clearly did not perform live with the kind of precision they exhibited in the studio, but their singing continued to be accurate and pleasant enough that people still buy these live recordings.

What makes this example even more impressive is that the Beatles were *singing*. Of all the forms of music making, singing may be the most contingent upon auditory feedback. In instrumental music, there is always some sort of visual feedback one can use, even if expert

Sound and Action in Music Performance
DOI: https://doi.org/10.1016/B978-0-12-809196-8.00002-4

performers usually do not. A guitarist can opt to look down at the fret board if need be, and a pianist can look at piano keys. By contrast, nobody can gaze at their vocal folds to check whether the appropriate level of tension has been achieved. And even if you could, which is possible with an imaging tool known as the *laryngoscope*, it is unlikely that the visualization could help. Unlike moving one's fingers, arms, and legs, the movements of internal muscles for pitch control eludes our conscious awareness.

For me, the fact that auditory feedback's importance is *qualified* makes this whole area of study so interesting. If performers relied entirely on auditory feedback we would be left with a very simple closed-loop system of the sort that was proposed more than 50 years ago (see Chapter 1). But instead we have a more nuanced system. Next, I explain the important theoretical implications that lay behind whether auditory feedback is important or unimportant.

WHAT DOES IT MEAN TO "NEED" AUDITORY FEEDBACK?

Some readers may be taken aback by my glib answer at the outset of this chapter, so let me frame it more clearly. The sense in which I use the word *need* is that feedback is a necessary component of retrieval in production. Without sound, according to this view, the performer may not be able to retrieve the serial order of events properly, correct errors when they come, and so on. The resulting performance might be a disaster, without the performer knowing it.

Though some may consider this proposal extreme, it is worth considering, insofar as some prominent historical figures have suggested such a view, the most famous figure being Watson (1920). The concept of an *associative chain* emerged naturally from the behaviorist paradigm he advocated. The production of an action leads to a sensory experience from the environment, which acts as a sensory trigger (one does not refer to "memory cues" in behaviorism) for the most strongly associated action, which should be the very next action in the sequence. Continue this chain of associations, and one has Rachmaninoff! Famously, Lashley (1951) pointed out some important problems with this account when it is applied to language. In language, the constituent units (phonemes, syllables, words) are combined in ways that elude a simple chaining account. Consider the sentence, "The *right*-wing speaker won her *right* to speak at the university, but mistakenly took a *right* instead of left turn at the intersection and did not enter the *right* building." It is hard to determine what should trigger retrieval for the word *right* in this context because there are three different words preceding it.

Furthermore, the word takes on different meanings as it appears in different locations. Thus, retrieval for the word *right* is probably not based on associating it to an immediately preceding word, but may instead reflect the use of a broader planning scheme that conceptualizes different instances of the word as a function of the sequence organized into a hierarchy of phrases. The repetition of individual items is even deeper in music, for which only 12 pitch categories (in Western music) can be combined into a vast number of possible combinations. Likewise, Lashley pointed out that the speed of music performance is probably too rapid for the feedback guidance that is necessary in associative chaining.

Nevertheless, the associative chaining idea holds some promise as an explanation. As remarked above, the associative chaining view would suggest that feedback absence should disrupt production. The serial chaining view has not diminished to the extent that it should be considered a "straw man" argument. Associative chaining may play an important role in memory retrieval during music performance (Chaffin, Demos, & Logan, 2016), as well as in statistical learning of sequential patterns in general, including language acquisition (Aslin & Newport, 2012). What is mostly at issue is whether such associative chains in memory are used to link perception and action sequentially to each other. In the most extreme version of this view (à la Watson), the performer may not be able to continue. A softer version of this hypothesis would predict that memory lapses during performance should increase when feedback-based retrieval cues are removed.

Thus, there is a reason to predict that auditory feedback might play a critical role in memory retrieval in performance, and thus may influence various dimensions of performance as a result, including the accuracy of pitch sequencing, temporal precision, and expressivity. I now consider research that has addressed whether this occurs. To give away the story a bit, the importance of auditory feedback is much more constrained than an associative chaining account would predict. During learning, auditory feedback plays a significant role. By contrast, after learning has occurred, the importance of auditory feedback diminishes considerably. Thus, I consider these two kinds of effects separately.

LEARNING WITHOUT AUDITORY FEEDBACK

A congenitally deaf individual will probably never learn to speak as fluently as a hearing child. This unfortunate fact illustrates how important it can be to have actions guided by sound during language acquisition. And yet things may be different for an instrument like the piano.

Unlike speech, where the control of articulation involves the rapid and precise adjustment of muscles that are hidden from view, one can learn a piece on the piano based solely on visual feedback and can slow down the tempo to facilitate learning. To some extent, after all, playing the piano can be reduced to a kind of button-pressing exercise.

As it turns out, pianists do poorly at learning to play a piece without hearing any auditory feedback, as demonstrated by Finney and Palmer (2003). When participants were asked to memorize unfamiliar notated pieces in silence, they made approximately 10% more errors in later recall than when they learned while hearing auditory feedback, whether or not they could hear themselves during recall. The difference in error rates between the two learning conditions is considerable; the absolute magnitude of this effect is comparable with, if not greater than, the disruptive effects of altering auditory feedback that will be discussed in later chapters. However, the removal of feedback did not destroy participants' ability to learn pieces entirely. Even when feedback was absent during learning, participants performed about 70% of the notes correctly, which is far greater than if participants were simply guessing (which would result in approximately 1 out of every 12 events being correct, or 8% accuracy). One possibility noted by the authors is that the presence of sound during learning facilitated deeper processing of learned melodies (cf. Craik & Lockhart, 1972).

Other studies pursued a related question: If a pianist needs to choose between practice with only movements (no sound), or only sound (no movements), which is better? Although most performers likely consider movement to be of paramount importance (it is movements that one focuses on first during skill acquisition) the ultimate goal of performance is to produce a sound.

The better choice seems to be learning via listening, though not by a large margin. A series of studies by Rachel Brown and Caroline Palmer at McGill University tested how auditory and motor information influence the recognition of learned melodies (2012) as well as recall via performance (2013). In the study concerning performance recall (most important for the present purpose), the authors found a considerable advantage of auditory-only over motor-only learning. For instance, recall of melodies during performance was about 7% more accurate when participants practiced by listening (and viewing notation) than when they practiced by playing melodies on a silent keyboard. This is a particularly compelling result given that no movement whatsoever occurred during auditory-only learning; participants had to learn movements based on imagining them. This difference may reflect the difficulty many participants have in imagining the sounds that may result from their actions, in contrast to imagining actions that may be used to generate a melody. An earlier study that used a similar methodology replicated the advantage of auditory-only learning, but only for participants with poor *aural skills*, referring to how well one can imagine the sounds that are associated with music notation (Highben & Palmer, 2004).

Another research group addressed the effectiveness of learning based on either auditory or visual feedback when one source of perceptual feedback is removed (Engel et al., 2012). Nonpianists learned melodies based only on auditory-motor mappings or on visual-motor mappings. A critical difference in this study from those reviewed in the previous paragraph is that both conditions here involved motor movements. What differed was the sort of feedback (auditory or visual) that one received from the motor movements. For the audio-motor condition, participants listened to and then reproduced melodies while never seeing their fingers or any form of visual notation. In contrast, for the visual-motor condition, participants watched videos of someone performing a melody (using a birds-eye view similar to observing one's own hand) and then performed it on a silent keyboard while viewing their own hands. The authors tested how well participants recognized melodies that were based either on an auditory presentation, or on a visual presentation. Participants were always accurate at recognizing melodies if the presentation modality matched the kind of feedback they had access to during practice (e.g., if participants learned melodies based on vision and then had to recognize a video of someone playing that melody on a keyboard). But what the authors were particularly interested in was *cross-modal transfer*: how well can you recognize a melody presented in one modality (e.g., sound) if you practice it using feedback from another modality (e.g., vision). The authors found an advantage for auditory-motor learning. When participants learned a melody with only auditory feedback they still performed well when recognizing the same melody based on visual information. By contrast, when participants learned a melody by watching a video of the performance, their recognition based on listening to the same melody was poorer. Thus, auditory learning generalized to vision more so than the reverse. As we will see later, this advantage for auditory learning was reflected in brain activity.

What about other instruments, and singing? I know of no studies that have manipulated feedback in comparable ways during learning. This is of course difficult because one cannot simply turn off the sound for a singer or instrumental performer. However, it is unlikely that the effect of removing feedback during learning would be any smaller for these kinds of conditions. If anything, the importance of feedback during learning may be greater for singing, given the demands of pitch control. Turning back to the example I used to start this chapter, note that the Beatles' success at performing without auditory feedback was only for pieces they learned and had performed together multiple times.

One can imagine much poorer performance for an initial practice session conducted live at the Hollywood Bowl!

To summarize, the presence of auditory feedback confers a significant advantage to learning. This may reflect the fact that the intended product of a performance—an auditory image (as argued in Chapter 1)—forms and consolidates during this time. However, learning does not depend entirely on the presence of sound, as a strict serial chaining view might predict. Moreover, practice based on motor and visual information alone may be superior to practice based only on auditory information. Later I reflect on the reasons why the effectiveness of learning varies in this way. In the next section, however, I turn to a logical next step for a performer: Having learned a piece of music, how dependent is subsequent performance on the presence of auditory feedback?

PERFORMING WITHOUT AUDITORY FEEDBACK

As we have seen, the presence of auditory feedback has a decisive effect on how accurately performers can learn and recall a new musical piece. Does the role of auditory feedback change when one performs a melody that has already been learned? During recall, does the sound of the notes one plays act as a retrieval cue, as suggested by early theorists like John Watson? Or, as proposed by Karl Lashley, does one form a memory schema that can run in a "feed-forward" mode and not rely on feedback? In general, it seems that Lashley was correct, particularly when it comes to the piano. Any effects of removing auditory feedback during the performance of a learned tune on the piano are subtle, if present at all. Stronger effects are found in singing, as one might expect, but still small in comparison with effects of altering auditory feedback (discussed in later chapters).

We can start with the aforementioned study by Finney and Palmer (2003). In addition to having pianists learn melodies with or without auditory feedback, the authors also had them recall melodies from long-term memory (from their own repertoire) with or without auditory feedback. One hypothesis at play was the classic *encoding specificity principle*, from memory research (Tulving & Thompson, 1973). According to this principle, recall improves if the environmental conditions associated with retrieval match those associated with encoding. The original demonstration of encoding specificity had to do with memorizing a list of words either on land or underwater; people could later recall the same list if recall happened in the same environment. If the same principle holds for music, then pianists who learn melodies with auditory feedback should recall them with greater accuracy if feedback is present, and also (more importantly) pianists who learn without auditory feedback should recall more accurately when feedback is absent. This is not what the authors found. Effects of feedback presence were due entirely to whether learning was conducted with auditory feedback. The presence of sound during recall had virtually no effect. In other words, people who practiced with sound had better recall regardless of whether the sound was switched on or off while they performed.

One possible rejoinder to this result has to do with the fact that Finney and Palmer had pianists learning melodies for the first time. What about the recall of melodies one has played for many years with auditory feedback? It seems as though sound may be dispensable in these cases too. In another study from the same paper, Finney and Palmer had pianists play pieces from their learned repertoire (i.e., recall from long-term memory) with or without auditory feedback; performance was similarly accurate in both cases. In an earlier study, Finney (1997) found no effect of removing feedback when pianists sight-read Bach two-part inventions.

A second rejoinder comes from the use of well-trained pianists. Perhaps this group differs from novices in their ability to use muscle memory in a way that frees them from the need to use auditory feedback. Not so. Pfordresher (2005) also found no effect of feedback removal when pianists or nonpianists performed simple melodies from memory on a keyboard. Of course, null results from traditional statistical tests are inherently ambiguous. A newer approach comes from Bayesian statistics, using an odds ratio known as the Bayes factor, which yields an odds ratio based on the degree of support for the null versus alternative hypotheses (Rouder, Speckman, Sun, Morey, & Iverson, 2009). When applying this measure to the normal and silent feedback conditions from Pfordresher (2005), it suggests "some evidence" for the null hypothesis: a ratio of 4.85 relative to the support for the alternative hypothesis.

Can we say, then, that auditory feedback plays no role in solo piano performance? Not quite. So far I have focused on studies that address the accuracy of recall in performance, that is, how many mistakes people make. Despite the transparent importance of accuracy for music performance, there are other more subtle factors that make a big difference for the listener. Accurate performances differ in quality based in part on the use of *expressive nuances* in performance. In Chapter 1, I discussed the expressive use of timing. Performers invoke similar changes in loudness, timbre, and articulation (e.g., *legato* vs *staccato*) as well. Perhaps auditory feedback influences a performer's ability to manipulate these expressive parameters.

Even here, effects are underwhelming. Repp (1999) analyzed expressive and nonexpressive (metronomic) performance of Chopin's Etude in E Major (op. 10, no. 3) with and without auditory feedback. Although there were statistically significant effects of feedback on timing, they primarily had to do with lengthening of the final interonset interval and did not suggest any reshaping of the overall structure. A later study tested the effect of feedback on expressive timing with more advanced measures of timing, called *dynamic time warping*, that measure the degree of similarity in relative timing across critical segments of a performance (Bishop, Bailes, & Dean, 2013). This study found no significant effects of auditory feedback presence on timing. Likewise, both of these studies looked at expressive changes in loudness (intensity) and found similar effects in each case to what was found for expressive timing. A later study by Laura Bishop showed no effects of feedback deprivation when changes to intensity were produced by manipulating a slider rather than by performing on a keyboard (Bishop, Bailes, & Dean, 2014).

At this point you may wonder if the use of piano in past research may bias the results. The piano has a discrete representation of pitch, with a one-to-one mapping of pitch to spatial targets (piano keys). The absence of auditory feedback may not be problematic in this case because the performer can focus on this clearly specified set of visuospatial targets instead, leading to intact recall of pitch sequencing. By contrast, singing involves the control of pitch on a continuous dimension, based on a complex set of muscles that are hidden from view. Might this make singing more susceptible to disruption when auditory feedback is absent?

It does. However, effects of masking feedback during singing are also not so strong as to suggest that singers "rely" on auditory feedback. One difficult methodological issue with singing is that singers receive self-feedback through the air and through bones of the head. The intensity of these summed feedback sources is high, which makes it hard to mask feedback at safe intensity levels. A particularly clever solution to this problem was reported by Beck, Rieser, and Erdemir (2017). These authors harnessed the fact that singing (or any other form of motor sequence production) is disrupted by a delay of auditory feedback (see Chapter 3, for more). Thus, if masking really works, this effect should vanish. The authors used responses to feedback delays to set an optimal masking level for each participant, who in their study included young children (aged 5–8), preteen and "tween" children (aged 9–12), and college-aged adults. All three groups were significantly disrupted by the absence of feedback while singing "The ABC Song." However, the magnitude of their mistuning was only about 8 cents on average. To classify as an error, deviations from sung pitch typically need to be at least 50 cents, which is half the distance between two adjacent pitches on a piano. Thus, although the effect of masking is important and significant statistically, we cannot say that participants made errors when feedback was absent. This stands in contrast to the large effects of removing feedback during learning, and alterations of feedback that will be discussed in later chapters.

Participants in Beck et al. (2017) had varying degrees of musical training. Other research has looked at trained singers, who are less disrupted by feedback masking than are untrained singers (Erdemir & Rieser, 2016; cf. Schultz-Coulon, 1978; Ward & Burns, 1978). In addition, according to the results of Erdemir and Rieser (2016) instrumentalists are just as strongly disrupted by feedback absence as nonmusicians while singing, even though instrumentalists were more accurate at singing in general. The ability to sing without auditory feedback therefore seems to rely on the formation of a motor plan based on vocal-motor training. More evidence for this claim was found by Mürbe, Friedemann, Hofmann, and Sundberg (2002, 2004). They compared effects of masking noise for trained singers during legato (i.e., sustaining each note) and staccato (making each note as brief as possible) performances. Masking effects did not differ across these conditions, despite the fact that much more auditory feedback is available during legato performance. The similarity in results suggests that auditory feedback may only play a role in guiding phonation at the beginning section of sung tones (present in both kinds of performance). Moreover, singers may rely for the most part on kinesthetic rather than auditory feedback.

IDEOMOTOR THEORY: WHY FEEDBACK IS NOT ALWAYS NECESSARY

Results to date thus suggest that auditory feedback may not be necessary for the production of learned action sequences. During learning, performers repeatedly experience associations between planned actions and perceptual feedback. As a result, with time the perceptual representation of a sound pattern and the actions one might use to create that same sound pattern form an integrated representation that may exist separately from either the perceptual representation or the motor representation on its own.

Although this kind of integration may not surprise practicing musicians, the broader implications of it for psychology are considerable. Theories advocating the integration of perception and action have come and gone during psychology's brief history. At its dawn, James (1890) famously advocated for the integration of perception and action via the *ideomotor* principle of action planning (although in point of fact, James was drawing largely on ideas from other scholars; see Stock & Stock, 2004, for a review). In James' own words, "...every representation of a movement awakens in some degree the actual movement which is its object..." (p. 526). By "representation" James means an abstract idea, or goal, associated with a movement. By this logic, simply forming the intention to perform an act ("I want to eat that sandwich") puts in motion the motor plans that would be used to realize that intention.

An important precursor to William James was the German scientist Herman Lotze. Whereas James focused on the triggering of actions by abstract ideas, Lotze (1852) emphasized the importance of activating *sensory* representations as the trigger for actions (for a comprehensive historical review, see Stock & Stock, 2004). Lotze's focus is more closely related to the topic of this book and, more important, offers a more easily testable hypothesis for motor control. For instance, it is hard to identify neural activity associated with "playing the C-major sonata by Mozart" at an abstract level, whereas activity that may be associated with listening to a piece of music is more easily addressed. Thus, one can evaluate whether neural activity associated with listening is sufficient to trigger muscle movements used to perform that music (i.e., finger movements).

As noted before, ideomotor theory faded from psychology for quite some time (see Hommel, 2009, for a review). Watson's behaviorist paradigm, which followed the early days of William James' prominence, advocated a focus on observable phenomena. Investigations into the association between perception and action thus would avoid hypothesizing about the kind of unobservable cognitive processes that may be responsible for forming these associations. In other words, according to behaviorism it simply did not matter whether there was a representation in the brain that combined perception and action. When cognitive psychology started to take over during the 1950s, psychology's focus turned toward the possibility that mental processes may function like computer programs. Ideomotor theory also did not fit well within this framework. Near the dawn of the 21st century, however, cognitive psychologists increasingly found an influence of actions and the body on perception and even higher cognition. Such results led to the idea of *embodied cognition* (Wilson, 2002; cf. Gibson, 1979), and a resurgence of research on ideomotor planning (for reviews, see Badets, Koch, & Philipp, 2016; Maes, 2016; Shin, Proctor, & Capaldi, 2010).

What kind of representation underlies the integration of perception with action? The ideomotor principle on its own does not answer this question specifically, and more than one possible answer exists, as illustrated in Fig. 2.1. It is possible that ideomotor planning may occur because of enhanced associations between existing representations for motor planning or perception. This possibility is illustrated in the upper right side of Fig. 2.1, in which learning enhances neural pathways between the auditory and primary motor cortices. Indeed this kind of learning was advocated strongly by Hebb (1949), who observed that associations based simply on temporal contiguity may be sufficient to

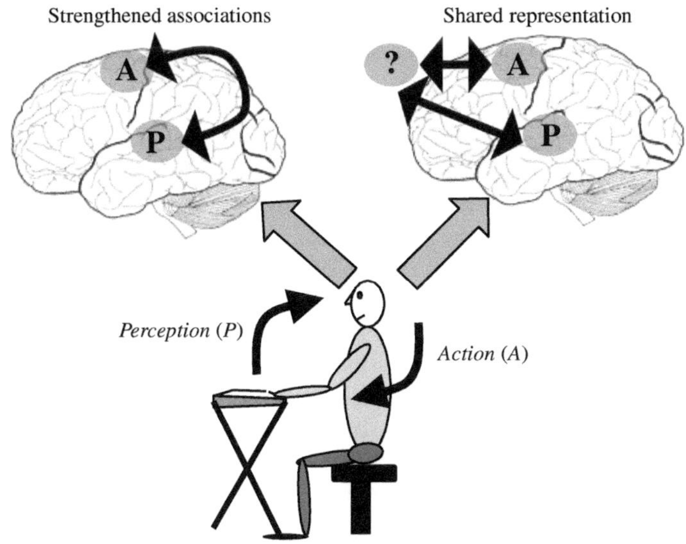

FIGURE 2.1 Two possible ways that musical experience performing a keyboard (lower) may facilitate associations between action planning (A) and perceptual feedback (P). Learning may enhance associations between representations used exclusively for perception or action (upper left), or may lead to the formation of shared multimodal representations (upper right).

cause learning to occur. Thus, the fact that every time you press a certain piano key the result is a specific pitch may cause neurons associated with both sides of this causal relationship to form an associated connection.

Alternatively, learning an instrument could cause new *shared representations* to form in which no division between perception and action exists. Several results, including many summarized in this book, point in this alternative direction. Variants of this idea exist in cognitive models of perception and action, such as the Theory of Event Coding, which was proposed by a group of researchers at the Max Planck Institute for Psychological Research, led by Hommel (2009, 2015), Hommel et al. (2001), and Prinz (1997). This theory proposes that shared perception/action representations are based on abstract codes, specifically, "...perceiving and planning are functionally equivalent, inasmuch as they are merely alternative ways of doing the same thing: internally representing *external events*" (Hommel et al., 2001, p. 860, italics added). Critically, the shared representation in this theory does not incorporate all the details of either perception or action, but instead represents only those features that characterize the external, or distal, realization. Thus, when playing a melody on the piano, the shared representation would code action planning with respect to keys on a keyboard—the goals of your actions—whereas details about muscle movements and finger selection would be reserved for the action-specific representation. Node Structure Theory (MacKay, 1987), discussed in Chapter 1, also proposes a shared perception/action network.

The shared representation construct helps explain why performers do not seem to need a melody's auditory feedback after learning to perform that melody (without auditory information). If a representational equivalence exists between action and perception, then any function taken on by auditory feedback may be substituted by action planning. The details of how such an architecture might work have been laid out in the Internal Models framework. A Forward Internal Model (described briefly in Chapter 1) adapts a motor command to constraints of the system that generates this command, as well as any learned distortions of outgoing information (e.g., perturbing forces placed on arm gestures, Wolpert et al., 1995). In this way, the model can predict the kind of sensory feedback the system ought to receive based on the predicted outcome of a planned action. Under normal feedback, the predicted output is compared with actual auditory feedback, and any discrepancies may lead to alterations of the forward model. This activity plays an important role during learning, hence the importance of auditory feedback presence at that stage. However, if auditory feedback vanishes, and the model has already been developed, the system can continue mostly unabated without feedback. The system can even correct errors when the model predicts that the motor command has not yielded the intended output, thus accounting for the fact that performers engage in corrective actions prior to making an error (in music, Repp, 1996; in speech, Levelt, 1983), which I review in Chapter 8.

There is another representation issue of importance. Ideomotor planning and forward modeling reference the anticipated perceptual output, not actual perceptions. How do we represent an "anticipated" perception? One possibility is mental imagery. A great deal of research on mental imagery suggests that the images are represented perceptually rather than propositionally (Pearson & Kosslyn, 2015). At the same time, neuroscience shows that primary senses often are not active during imagery (Zatorre & Halpern, 2005). Thus, these representations may be abstract in a way that is similar to what is proposed in the Theory of Event Coding. Moreover, some evidence from neuroscience suggests motor activations for auditory imagery in planning areas, as I will discuss in the next section, which in turn suggests that auditory imagery may prime motor planning. This also matches claims of the Theory of Event Coding (TEC) in that imagery may serve as a shared representation for perception and action. Keller (2008, 2012) picks up on this in using the term "musical imagery" to refer to a multimodal perception/action representation used by musicians.

NEURAL MECHANISMS FOR IDEOMOTOR ASSOCIATIONS

If learning leads to distinct shared representations for perception and action (as in Fig. 2.1, right) then one may expect to find networks in the brain that respond integratively to perception and production tasks. Several studies have measured the formation of neural auditory-motor associations in music learning using *functional magnetic resonance imaging*, or fMRI, a leading technique for localizing brain activity (though not the timing of brain responses) that was described in Chapter 1.

It can be difficult to conduct music performance experiments using fMRI. As in an anatomical MRI, fMRI images are collected while the participant lies supine (i.e., on one's back). Thus, a pianist must perform in a nonstandard position, often with a keyboard suspended above the waist. Second, the MRI machine is effectively a large tube, and one's head and shoulders need to rest inside the tube while measurements are made. This can lead to claustrophobia in some participants, and prevents all participants from looking down at the hands during performances. Finally, because of the intense magnet in the MRI, specially designed keyboards with no metal must be used.

Several researchers have risen to the challenge and used fMRI to address how musical training leads to audio-motor sensations as predicted by the ideomotor planning hypothesis. Three studies have used longitudinal designs to measures changes in brain activity both before and after musical training on a piano keyboard (Chen, Rae, & Watkins, 2012; Lahav, Saltzman, & Schlaug, 2007; Mutschler et al., 2007). In these studies, nonpianists were brought to the laboratory over a span of time that could take up one session (Chen et al., 2012; Mutschler et al., 2007), or be distributed across 5 days (Lahav et al., 2007). Participants learned to play melodies on a keyboard while hearing normal auditory feedback. A critical issue in all these studies was how the perception/action associations that occurred during training influenced brain responses to perceived music. The researchers measured the effect of learning by comparing brain responses with melodies that the participants learned versus other melodies (Lahav et al., 2007; Mutschler et al., 2007), or by tracking changes to brain responses during music perception across training blocks (Chen et al., 2012).

All three of these learning studies revealed brain activity in two motor-related areas during music listening tasks highlighted in Fig. 2.2. The premotor and supplementary motor areas both contribute to motor planning (Zatorre, Chen, & Penhune, 2007) and to synchronization of motor and auditory activities in the brain (Herholz & Zatorre, 2012). The other area is the inferior frontal gyrus, otherwise known as "Broca's

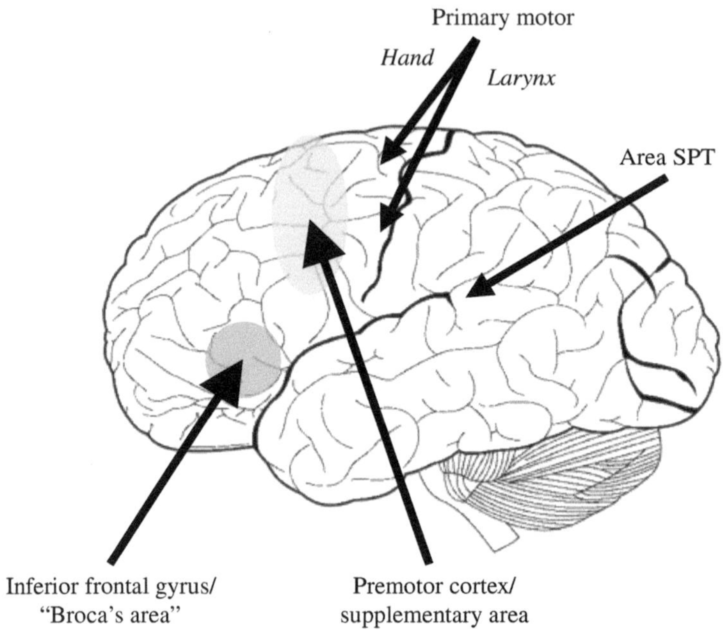

FIGURE 2.2 Approximate location of brain areas exhibiting motor responses during music perception after training in four longitudinal studies (Bangert et al., 2005; Baumann et al., 2007; Chen et al., 2012; Lahav et al., 2007; Mutschler et al, 2007). Darkness reflects agreement of location across studies, which was greater for inferior frontal gyrus, which was also found in a study involving nonmusical auditory-motor associations (McNamara et al., 2008). Other relevant brain areas are highlighted with arrows.

area," which was historically associated with speech production but now is known to have a more general role, including processing of musical structure (e.g., Koelsch, Gunter, von Cramon, Zysset, Lohmann, & Friederici, 2002; Kunert, Willems, Casasanto, Patel, & Hagoort, 2015; Maess, Koelsch, Gunter, & Friederici, 2001; Patel, Iversen, Wassenaar, & Hagoort, 2008). Broca's area technically exists only in the hemisphere that is dominant for language (usually the left), but in musical tasks one typically finds activity in the same area of the brain on the alternate side (usually the right), referred to as the *homolog* to Broca's area. Note that neither of these areas is responsible for direct control of hand movements; that kind of control occurs within the primary motor cortex, which is right around the middle of the brain's surface. The absence of activity in the primary motor cortex suggests that motor associations reflect more abstract planning stages as opposed to the actual preparation of motor movements (cf. Fig. 1.1). This result is in keeping with the nature of motor representations thought to form a "common code" with perception (Hommel et al., 2001), and also converges with motor activations found during auditory imagery (e.g., Leaver, Van Lare, Zielinski, Halpern, & Rauschecker, 2009; Lima et al., 2015).

As I have noted before, studies of performance have typically been limited to the keyboard, as in the studies summarized in Fig. 2.1. However, a recent study broadened our perspective by addressing how training on the cello influences auditory responses to scales and melodies that participants had learned to play (Wollman, Penhune, Segado, Carpentier, & Zatorre, in press). The authors used a specially designed cello for use in the scanner, with no metal parts and without the large wooden body of a standard cello. Unlike other stringed instruments like the violin and guitar, the orientation of the cello relative to one's body makes this instrument more suitable for performance in the scanner. As in studies of piano, practice on the cello increased responses in the supplementary and premotor areas during listening tasks, in addition to parietal activations that were more superior (toward the top of the head) than in piano studies. Activations in the inferior frontal gyrus did not increase with training, although activations in an adjacent area under the surface—the middle frontal gyrus—did.

The auditory-motor associations found in these studies may reflect a broader system for sensorimotor integration. An additional study by McNamara et al. (2008) measured brain responses to synthesized tones that are associated with meaningless hand gestures (e.g., fanning out the fingers and positioning the thumb over the ring finger). The key area whose response changed over time was the inferior frontal gyrus, shown in Fig. 2.2. However, this experiment did not yield activity in the premotor and supplementary motor region, perhaps because participants did not generate sounds via hand gestures on their own.

The inferior frontal gyrus is significant because it is part of a network that has been dubbed the *mirror neuron system*. Mirror neurons were first discovered by a happy accident in a research laboratory studying the neurophysiology of motor planning in the monkey cortex (di Pellegrino, Fadiga, Fogassi, Gallese, & Rizzolatti, 1992). A member of the laboratory noticed that a neuron that fired as one of the monkeys reached toward a food pellet also fired while observing the assistant placing the food pellet in front of the monkey. Over the course of several studies, the research team proposed that the cortex possesses a population of neurons that are not exclusively perceptual or motor, but instead serve a dual purpose. The analogy of a mirror was chosen because these neurons treat action observation similarly to watching yourself perform that action in a mirror. Whereas the original discovery was based on single-cell recording, and similar single-cell responses have been found in humans (Mukamel, Ekstrom, Kaplan, Iacoboni, & Fried, 2010), most studies of this sort on humans have been done with fMRI—a much grosser measure of activity (Logothetis, 2008). Although there has been a great deal of controversy surrounding the role and even the existence of mirror neurons in humans (for a lively discussion, see Gallese, Gernsbacher, Heyes, Hickok, & Iacoboni, 2011), these studies minimally show that a sharp division between perceptual and motor processing in the brain may not exist and that at least some areas of the brain may serve a dual purpose.

Another neuroimaging study I discussed earlier was the study of cross-modal transfer by Engel et al. (2012), where nonmusicians learned to perform melodies on a keyboard while experiencing either auditory or visual feedback but not both. Like behavioral results discussed earlier, brain activations suggested greater transfer from auditory learning to vision than vice versa. The relevant brain activity was also found in the inferior frontal gyrus, thus expanding on its role in forming associations during sensorimotor learning. Specifically, activations in this area were greater when people observed silent video clips of people performing melodies they had learned with auditory feedback, than when they observed clips of people performing unlearned melodies. Thus, even though participants in these conditions did not see themselves play, the conjunction of auditory feedback with motor movements led to activity in the inferior frontal gyrus that facilitated visual imagery for the movements. Thus, this brain area may play a role in forming both sensorimotor associations and also associations across different sensory modalities.

A potential limitation of these ambitious studies that use longitudinal learning designs is that the span of time used for training inevitably falls short of the accumulated experience found in expert pianists (cf. Ericsson, Krampe, & Tesch-Römer, 1993). Other studies have focused on expert performers to determine the kind of auditory-motor associations that arise after years of training. Activation in the areas highlighted in Fig. 2.2 is greater among pianists than nonpianists during auditory perception tasks (Bangert et al., 2005). Within pianists, these areas are more active while listening to music than during rest (Baumann et al., 2007). Expert violinists also show greater activity in these areas while imagining performance than do novice violinists (Lotze, Scheler, Tan, Braun, & Birbaumer, 2003). Two other studies yielded slightly different results, however, possibly due to subtle differences in measures and techniques. For instance, an earlier study in which pianists detected out-of-key notes in familiar classical melodies yielded activations in the hand area of the primary motor cortex that were specific enough that the activations were finger-specific (Haueisen & Knösche, 2001). This study used magnetoencephalography (MEG), which has slightly lower spatial resolution than fMRI and requires the use of computer modeling to locate areas. Nevertheless, it is noteworthy that this study found activation specific to motor movements while other studies have not. For reference, Fig. 2.2 highlights the area of the primary motor cortex that controls hand movements, as well as an area recently identified as controlling vocal pitch (Brown, Ngan, & Liotti, 2008).

Another study focused on an area of the brain that serves a sensorimotor function for speech. This area sits at the junction of the Sylvian fissure (for speech), the parietal lobe, and the temporal lobe, and is thus called area SPT (Hickok, Buchsbaum, Humphries, & Muftuler, 2003). Brain activity in this area (via fMRI) has been found when people imagine speaking a sentence or humming a tune, thus suggesting that it plays a role in both music and speech. However, this area exhibited significantly less activation when participants imagined playing the melody on a keyboard (Pa & Hickok, 2008). Thus, sensorimotor associations for the voice may rely on a different network than the network that is used for keyboard performance.

Does sensorimotor learning fine-tune the timing of neural responses to a perceived melody? As discussed in Chapter 1, the timing of brain activity is best measured through techniques like EEG and ERP. A recent study addressed how learning affects the timing of brain responses (Mathias, Palmer, Perrin, & Tillmann, 2015). A sample of pianists learned melodies via listening, or by performing them. After this learning phase, pianists' brain activity was measured while listening to melodies that could contain altered pitches. Brain responses were greater when altered pitches were present in melodies that were performed during learning, as opposed to melodies that were learned only via listening. These neural responses occurred rapidly, within 200 ms of the change, which suggests that sensorimotor associations enhance the tuning of neural response to pitch (cf. Trainor, Shahin, & Roberts, 2009). Although it is difficult to localize the EEG signal, results of a computer model suggested that the responses originated in the premotor and supplementary motor areas.

Though using some different techniques, all studies so far have used neuroimaging, a method of measuring the brain activity that accompanies some kind of a task. A limitation of neuroimaging is that it does not establish the *necessity* of some brain activity to a task, nor does it suggest that the brain activity *causes* the behavior to occur. A good technique for establishing this causality and necessity is the use of *transcranial magnetic stimulation* (TMS), which can be used to stimulate or disrupt brain activity. A team of researchers in Germany used TMS to examine how music listening influences motor responses (D'Ausilio, Altenmüller, Olivetti Belardinelli, & Lotze, 2006). In this study, the researchers stimulated the right motor cortex of amateur pianists who were listening to either a Bach keyboard piece or a flute piece. The first time this happened, both pieces of music were novel to them. Then the pianists would practice the Bach piece and experience TMS again.

Practicing Bach led to greater motor excitability during the listening task: TMS would cause slight muscle contractions in the forearm, as if the music were compelling the participant to move his or her fingers! Similar changes did not occur for repeated exposures to the flute piece. This work complements earlier research using TMS that demonstrated that mental or physical practice can enhance the responsiveness of motor commands (Pascual-Leone et al., 1995). These effects are enhanced even further if you listen to a melody that you know how to play.

INDIVIDUAL DIFFERENCES IN IDEOMOTOR ASSOCIATIONS: THE CASE OF POOR-PITCH SINGING

An important prediction of ideomotor models, including the Theory of Event Coding, is that perception/action associations should be bidirectional. Thus, although the main focus of this chapter is on how action planning elicits imagery associated with anticipated consequences of those actions, ideomotor planning also predicts that perceiving or imagining a stimulus should prime planning in motor areas. Indeed, this has been the main focus of most studies of ideomotor planning. For Internal Models, this kind of activity relies on an Inverse Internal Model, which uses a target perceptual state to model the actions that should yield that state. These ideomotor associations are also critical to the H-PAC model proposed in this book, which will be detailed in the next three chapters.

In music performance, recent research has focused on singing as a way to measure individual differences in this inverse use of perception/action associations. Singing usually involves the imitation of a perceived target, unlike the piano where such "performing by ear" is less common (and actually quite difficult relative to notation; see Engel et al., 2012). Although this kind of "call-and-response" activity for singing feels natural to most of us, it is worth considering the intricacies of this task. To match pitches in a melody that you've just heard, one has to translate the pattern of sounds into a series of muscle states in the larynx. Despite all this, a sizeable proportion of adults are able to match sung pitches within a semitone of the target (Berkowska & Dalla Bella, 2013; Dalla Bella et al., 2007; Hutchins & Peretz, 2012a; Pfordresher & Brown, 2007; Pfordresher & Larrouy-Maestri, 2015). Among the rest, to what degree is their deficit a failure in ideomotor planning?

As it turns out, ideomotor planning, or the use of an Inverse Internal Model, may constitute the primary source of singing accuracy. Part of this conclusion surfaces by elimination. Although perceptual capacities play some role in singing accuracy, and people suffering from congenital amusia (a pitch processing disorder) often sing poorly (Dalla Bella, Giguère, & Peretz, 2009; Hutchins, Zarate, Zatorre, & Peretz, 2010), perception deficits may not be necessary for inaccurate singing (Pfordresher & Brown, 2007). Likewise, although inaccurate singers exhibit accurate pitch matching with other motor systems (Hutchins & Peretz, 2012a, 2012b), singing accuracy may not be reducible to the ability to control laryngeal muscles (Pfordresher & Mantell, 2009). Finally, inaccurate singing may not simply reflect memory capacity, given that individual differences are found even for the matching of single pitches, for which no memory load exists. Based on these results, a deficiency of ideomotor planning appears to be the only source of deficiency remaining.

Two sources of positive evidence also support the idea that singing accuracy relies on an ideomotor planning or inverse model mechanism. First, inaccurate singers appear to form less vivid auditory images than more accurate singers (Greenspon, Pfordresher, & Halpern, 2017; Pfordresher & Halpern, 2013). If, as suggested earlier, imagery constitutes the form of a shared representation, reduced vividness for imagery suggests that these representations may be ill-formed. A second finding focused specifically on an implication of the Inverse Internal Model construct (Pfordresher & Mantell, 2014). One role of such a system is that it allows individuals to mimic motor gestures that they have not produced before. If I hear someone sing "Doe, a Deer" for the first time, the internal model allows me to simulate the motor processes that would lead to that song because the model provides an abstract framework that generalizes across a variety of gestures. If, on the other hand, the model does not serve this function, then the ability to imitate a novel sequence should suffer. However, the ability to replicate what one has already produced might remain unimpaired and rely on muscle memory. This is what was found. People in general do better at imitating recordings of themselves singing (Hutchins & Peretz, 2012a). This extends to an inaccurate singer, who will do a better job at imitating their own out-of-tune performance of "Doe, a Deer," than another person's accurate rendition, or even an out-of-tune rendition from another inaccurate singer. However, this advantage is significantly larger for inaccurate singers than it is for accurate singers.

Why might ideomotor planning fail to work properly for inaccurate signers? We do not know the answer at present, but one possibility is that inaccurate singers do not fully explore the pitch range of their voice while singing (cf. Rutkowski, 2015). As a result, the mapping between perception and action for these individuals may become constrained and inflexible, leading to patterns of bias like those observed here, along with the failure to form vivid auditory images based on sensorimotor interactions (Pfordresher, Halpern, & Greenspon, 2015).

CONCLUDING REMARKS

Although the presence of auditory feedback can facilitate performance, most effects are surprisingly subtle. For piano, effects of feedback presence are limited to early learning, and possibly the reliability of certain expressive nuances. Stronger effects are found for singing, with masking of one's voice leading to mistuning. However, these effects pale in comparison with the effects of altering auditory feedback, which I discuss in the next two chapters. Ultimately the interesting question arising from the present chapter is why the role of auditory feedback's presence is so limited.

Because the brain automatically generates a kind of auditory representation for the sound sequence one intends to produce, auditory feedback becomes optional. Thus, we can have the Beatles playing successfully to a deafening crowd, just as we have participants in experiments where the researcher chooses to turn the speaker off on the electric keyboard. There are, as some readers will know, other effects of these learned associations having to do with altering key-to-pitch mapping relationships on the keyboard. I will discuss these effects in Chapter 4. Before that discussion, however, I will turn to some of the most robust effects of altered feedback to date: effects of altering synchronization between perception and action.

CHAPTER

3

Binding Perception and Action in Time

"The Edge," lead guitarist for the rock band "U2," is famous for his unique performing style. Rather than perform virtuoso solos, The Edge specializes in a percussive kind of style that harnesses the use of an effect pedal that produces a kind of artificial reverberation, called the digital delay effect. When this delay is timed in the right way, one can strum on every beat and hear both the immediate sound of that note or chord as well as a repetition of the event on every offbeat. It sounds a bit like Philip Glass for the rock world (take a listen to U2 songs like "Bad" and "Where the Streets Have No Name" for examples).

The Edge's signature style is worth recognition for its ingenuity, and also because it is a lot harder to perform than you might think! I performed in several rock bands during high school and college, and worked with many excellent guitarists. All of them could issue a rapid sequence of notes in a guitar solo, but only one or two could master The Edge's style. Why? Because The Edge effectively overcame one of the most disruptive alterations of auditory feedback on record. His use of the digital delay closely resembles the paradigm known as *delayed auditory feedback* (DAF), which will dominate this chapter.[1]

This chapter reviews critical research having to do with the role of temporal synchrony between auditory perception and action. DAF, in most cases, is disruptive because it disrupts the natural synchrony we come to expect. This research has a long history, and as such I begin by reviewing some classic work that uses DAF, much of which involves speech but with considerable relevance for music performance. I then go on to discuss more recent research concerning the role of temporal synchrony both in simple tapping tasks and more complex tasks

[1]An audience member alerted me to this connection after I gave a talk on auditory feedback at McMaster University in 2011.

Sound and Action in Music Performance
DOI: https://doi.org/10.1016/B978-0-12-809196-8.00003-6

involving music and speech production. Finally, I reflect on the theoretical implications of these findings for how perception and action are coordinated.

The importance of feedback timing is not a foregone conclusion. Since Helmholtz, we have known that neural conduction times vary across different feedback sources. More recent work suggests that it takes a considerable amount of time to register auditory feedback consciously. Furthermore, musicians have to control action while performing alongside other musicians. The temporal pattern associated with this collective activity can be complex and multilayered. From this kind of evidence one might guess that our brains were set up to be able to disengage from auditory feedback, ignoring it during situations when the timing may conflict with what we hope to do. The research I summarize here shows that this is not the case.

AUDITORY FEEDBACK MUST BE SYNCHRONIZED WITH ACTIONS: EFFECTS OF DELAYED AUDITORY FEEDBACK

The first empirical evidence for the importance of having auditory feedback synchronized with one's actions arose in the early 1950s, through experiments concerning the effect of DAF on speech production. Two laboratories discovered similar results independently of each other during the summer of 1950 (Black, 1951; Lee, 1950, 1951). In each case, researchers configured a tape recorder to record and play back simultaneously, then separated the recording head from the playback head to create a delay that varied as a function of their separation. This manipulation was inspired by the surprising difficulty soldiers faced during World War II when they had to fire guns and missiles with a considerable delay between initial firing and the point of contact.

Lee (1950, 1951), working out of an army research laboratory, found that delays of approximately 250 ms would slow down speech considerably, but if speakers fought against this tendency they started to repeat syllables uncontrollably, in a kind of artificially induced stutter. Interestingly, these two distinct reactions suggest the use of different strategies in response to DAF. At Ohio State University, Black (1951) ran an experiment with more participants and delay conditions, reporting significant effects on speaking rate (slower) and loudness (greater) when people spoke with DAF.

A striking feature of this effect is its robustness and reliability. DAF, under the right circumstances, can incapacitate a fluent speaking adult, causing speech to take on the characteristics of pathological stuttering or slurred speech. The effect is replicable enough that it can be observed within an individual at the first exposure. As noted in Chapter 1, I sometimes fall prey to the DAF effect from a delay produced by my office phone. And yet, despite its long history, the basis for this striking effect is still not fully understood, although I will offer a proposed account in this chapter that I hope provides a first approximation.

Effects in speech are well and good, but this book primarily concerns music. Music, like speech, is a rhythmic auditory signal. However, the rhythms in music are more stable than in speech. Furthermore, instrumental musicians need to learn new associations between action and perception and thus may be adaptable to alterations of feedback in a way that speakers in general are not. Thus, it is not a foregone conclusion that effects in speech should be replicated in the domain of music (see Chapter 7 for further discussion).

In this case, however, music and speech converge closely. The first formal published study of DAF effects in music was by the music educator Larry Havlicek from the University of Kansas, publishing his findings nearly 20 years after the first studies of DAF in speech (1968), though anecdotal comments about DAF effects in music were reported earlier (Kalmus, Denes, & Fry, 1955). Havlicek (1968) had music education students with considerable training sight-read two pieces with normal auditory feedback and then two other pieces with a feedback delay of 200 ms. What is considerably impressive, even by today's standards, is that the author analyzed performances on a variety of musical instruments: piano, strings, brass, and woodwind. By contrast, most music performance studies in recent years simply include the piano. Havlicek found significant increases in performance errors for performances with DAF, relative to normal feedback, with no difference across instruments. This initial study demonstrates the generality of DAF across different forms of motor production.

And so a flurry of activity in DAF was seen in the 1950s and 1960s, at which time progress seemed to have run its course and activity largely stopped. The paradigm then resurfaced in the 1980s when Peter Howell and Donald MacKay introduced new interpretations of the effect that were introduced in Chapter 1 and that are presented in more detail below.

THE CRITICAL IMPORTANCE OF THE CRITICAL DELAY

The most theoretically important feature of DAF's effect may not be the fact that it disrupts performance, but that different delay lengths have different effects. Not all delays disrupt performance. Consider the very short-latency delays you may experience in a large cathedral.

Thankfully, delays associated with room reverberation, partly by virtue of their short latency (Beranek, 2004), do not disrupt people like DAF does. Likewise, a very long delay, though more disruptive than reverberation, may not be as detrimental to production as a delay of the magnitude used in many studies. Other, more intermediate- length delays, by contrast can be highly disruptive though to differing degrees. Thus, the degree of disruption caused by a given delay length may tell us something about how perception and action interact. Of central importance is the notion of a *critical delay*: the delay length that wreaks the greatest havoc on the performer (MacKay, 1987).

The very first published report of DAF in fact did not find a critical delay (Lee, 1950), a result that reflected the limited range of conditions the author used. This study thus stands as an example for why it is important to explore a phenomenon exhaustively before theory development. Lee found that the degree of slowing varied linearly with the magnitude of the delay, up to a maximum of 280 ms, which also happened to be the longest delay he investigated. Lee saw these effects as being consistent with a speech-monitoring approach in which each produced element of speech must be confirmed through feedback for production to continue. We can identify this as a strong example of a feedback-control theory discussed in Chapter 1. Lee went on to develop a quantitative model of the DAF effect. According to his model, any lengthening of the time between a produced speech element and feedback from that element should elongate speaking times. In fact, he proposed an equation to predict this in his 1950 paper:

$$T = N \times (D + B) \tag{3.1}$$

where T refers to the total amount of time it takes to produce a spoken sequence (list of syllables, text, etc.), N refers to the number of elements produced (e.g., syllables), D refers to the length of the delay, and B refers to the average time it takes for speech under normal baseline conditions per each unit. It has been estimated that syllables in English average about 200 ms each. Thus, if someone were to produce a sequence of 10 syllables, the total duration should be about 2000 ms (10×200). If we add a feedback delay of 250 ms, it becomes $10 \times (250 + 200)$ or 4500 ms. This is pretty close to what Lee (1950) found for speech production with DAF.

Lee deserves substantial credit for developing a concrete, testable theory of DAF based on systematic assumptions about the underlying architecture. At the same time, this model suffers a deep flaw that could have been avoided by a more exhaustive experimental design. Eq. (3.1) predicts an ever-increasing DAF effect. The longer the delay, the slower the performance—there is no critical delay. For instance, this model predicts that a 500-ms delay should lead to a total duration of 9000 ms of the aforementioned 10-syllable sequence. This prediction does not match human behavior.

In fact, a problem with Lee's original account surfaced from the results of Black (1951), who used a wider range of delays (10 conditions, with delays ranging from 30 to 300 ms and a no-delay baseline condition). Black also found increasingly slow production as delay amounts increased, but only up to a delay of 180 ms. For delays longer than this, the effect of DAF diminished, leading to a nonlinear effect. A later study by Fairbanks and Guttman (1958) included an even broader range of delays, up to 800 ms. For delays up to 200 ms, errors increased and timing slowed, but beyond that length the trend reversed as delays got still longer.

Fig. 3.1 shows data from four early studies of DAF, selected because they included a wide range of delay values. In addition to the three studies of speech described above this figure includes data from the first study of music performance to explore effects of DAF across a range of delays, in which trained keyboardists performed Bach minuets (Gates, Bradshaw, & Nettleton, 1974). Conveniently, all studies except for Lee (1950) included tables with summary statistics for each condition; data from Lee (1950) were estimated based on Fig. 1 of his paper. I represented disruption here as the percent of change in performance rate relative to a baseline condition with no delay. This allows a more uniform representation of effect sizes across the tasks, which differed in the materials and measures used (e.g., Lee presented total reading time,

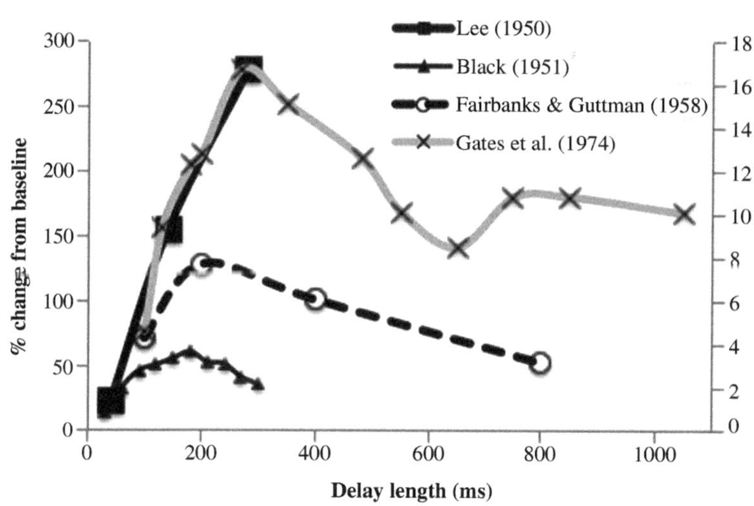

FIGURE 3.1 Results from four early studies of DAF. Means from Gates et al. (1974) are shown on the right-side y-axis.

whereas Fairbanks and Guttmann presented rate per syllable). These studies show a rough correspondence in the length of the critical delay, despite having many differences across studies. One noteworthy difference is that effects of DAF on music performance (Gates et al., 1974) were much smaller than the other three studies, which focused on speech. Thus, the effect of DAF on music is plotted on a secondary *y*-axis. Some have claimed that music performance exhibits a critical delay at a longer interval than found for speech (e.g., Howell et al., 1983), although as can be seen it is possible that the speech data of Lee (1950) show a similar peak.

Thus, we have early evidence that the function relating delay length to performance is nonlinear, with a critical delay in the vicinity of 200 ms. The deeper question, of course, is why this nonlinearity occurs. We now consider two alternative accounts, and how production rate (tempo, in musical contexts) plays a critical role. At issue is whether perception and action are temporally associated in *absolute* or *relative* time. According to an absolute time framework, one associates action and perception based on fixed time intervals, like a stopwatch. So regardless of whether you play a Bach invention at a tempo of 100 or 600 beats per minute (bpm), a 200-ms delay may disrupt your performance in comparison with other delay lengths. According to a relative time framework, perception and action are associated by virtue of *rhythmic phase relationships.* As a result, the delay causing maximal impairment will depend on rate or tempo (these terms mean the same thing, but tempo is associated with music). I will discuss these perspectives by contrasting two major theories of perception and action that arose in the 1980s.

Absolute Time and Node Structure Theory

We first consider the absolute time framework. Probably the most systematic account that follows from this assumption comes from Node Structure Theory, which was introduced in Chapter 1 (MacKay, 1987). This theory proposes that perception and action rely on a common set of interconnected "nodes," which are loosely analogous to neurons. As in the nervous system, activation within the node network depends in large degree on dynamic fluctuations in the sensitivity of content nodes (i.e., nodes designed to select a specific event for production or perception) to input. And these dynamic fluctuations are central to the theory's account of disruption from DAF. The resulting "node-recovery cycle" is illustrated in Fig. 3.2, and is segmented into three phases. In the first phase, input to a node causes priming of that node to increase. If priming reaches a critical level then the node becomes activated (the second

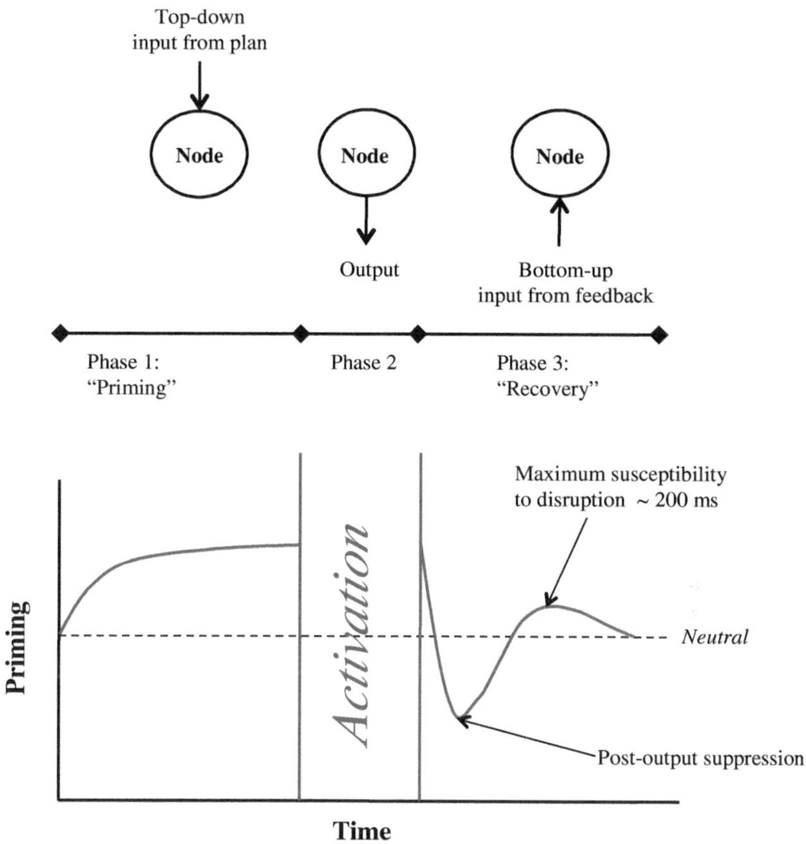

FIGURE 3.2 Illustration of Node Structure Theory's predictions for DAF effects, based on the node-recovery cycle, shown in the lower panel. This cycle is divided into three phases, as illustrated by the middle bar, and relates to the types of processing done within a single content node, which is illustrated at the top.

phase). Activation means that the node performs its designated role, which may be producing a musical note. This is akin to an action potential in a neuron. After this activation phase, there is a third "recovery" phase where the resting level of node priming (independent of input) varies. At the start of the recovery phase the node enters a *refractory period* when its resting level dips below baseline. As a result, it would take more input for the node to fire during this period. Next there is a brief rebound during which the node exhibits *hyperexcitability*: the resting level goes above baseline, causing the node to become more susceptible to activation. Note that all these stages are directly analogous to the dynamics of a neuron in the central nervous system.

Recall that nodes in this theory are bidirectional. As such, the node associated with producing a note or syllable may be activated by input from one's intentions to produce it (called *top-down* input), or may be triggered by external input including auditory feedback (*bottom-up* input; it is worth noting that these nodes also serve for strictly perceptual tasks, not just as a way of processing auditory feedback). The refractory period is critical here. If auditory feedback were to prime a node used both for perception and production, the system may repeatedly produce the same word (as a stutterer might do). The refractory period prevents nodes from responding to feedback, which follows activation of the node closely in time.

This is where DAF effects come in. According to MacKay, DAF disruption occurs when the timing of the delayed feedback's onset coincides with the hyperexcitable point in the node's recovery cycle. When timed in this way, DAF can easily reactivate the node, leading to repetitions of the event and the "artificial stutter" noted by Lee (1951). In other words, the critical delay found in DAF studies reveals the timing of this hyperexcitable point. Although the theory does not say so explicitly, one would presume that slowing of production by DAF may be an indirect consequence of this reactivation. Importantly, Node Structure Theory commits to an absolute time basis for the critical delay. None of the phases in the node-recovery cycle are influenced by higher-order temporal patterns.

If this account were true, then associations between perception and action should not be influenced by production rate. Binding of perception and action instead should be based on a fixed time window of about 200 ms, outside of which perception can disrupt action planning. Support for this idea has been found in various tasks. For instance, although people make more errors when speaking at a fast than at a slow rate (this is the classic *speed/accuracy tradeoff*), the effect of DAF was not found to be any different across tempi when people produced speech at a slow versus a fast rate (cf. Butler & Galloway, 1957; MacKay, 1968, 1970; Siegel, Fehst, Garber, & Pick, 1980), and an early study suggests that DAF effects are consistent across sentences that contain words with longer syllable durations in comparison with shorter syllable durations (Huggins, 1968). At the same time, other factors that are independent of production rate do influence which delay is most disruptive. For instance, although the rate at which one presently speaks does not influence how disruptive a given delay will be, the *fastest possible* rate at which one can speak does have an effect. People who are able to speak faster are disrupted by relatively short delays regardless of how fast or slow they speak while experiencing DAF. As a result, adults may be most disrupted by a delay of 200 ms, whereas a young child may be most disrupted by a delay of 600 ms (MacKay, 1968;

Siegel et al., 1980). According to the theory, this difference reflects how quickly nodes recover after an action is generated—when your node recovers more quickly (which is associated with greater fluency in general), you may be disrupted by relatively short delays. Yet within an individual the recovery cycle does not vary with production rate, and is therefore still based on absolute time.

There are two other constraints on this theory that have to do with *feedback content*, which refers to what kind of event category feedback represents. Although DAF disruption in this theory is based on absolute time, the content of feedback also matters.

First, DAF effects in this theory can only occur if the content of the delayed feedback matches the content of what you planned and produced. For instance, if you produce the syllable "glass," and feedback is delayed you should only be disrupted if the feedback still sounds like the word "glass" that you spoke. That is because it is the node associated with "glass" that is undergoing the dynamic changes shown in Fig. 3.2. However, if the sound is changed so that "glass" now sounds like "house," then you should not be disrupted because the node for "house" has not entered a hyperexcitable phase. The question of whether this constraint holds will be a major issue taken up in Chapter 4.

Second, this theory predicts that DAF effects will depend on how variable or complex the content of the entire sequence is. In an unpublished study by MacKay and Burke (1972, cited in MacKay, 1987), participants either spoke full sentences, like most other studies reviewed so far, or had participants repeat a single syllable over and over. Disruptive DAF effects were only found for sentence production, and not for repeated syllables. MacKay (1987) explains this phenomenon as a byproduct of practice (i.e., repetition), which may speed up the recovery cycle of a node. This constraint actually poses a methodological problem that we will take up later in the present chapter.

As we will see, the support for many of these critical predictions is mixed, particularly when it comes to musical behaviors (which admittedly was not the primary goal of MacKay's theory, though it purports to be general enough to apply to music as well as speech). Nevertheless I will argue that some loosening of its original assumptions, particularly the role of absolute time, make the remaining aspects of the theory highly useful as an account of sensorimotor associations.

Relative Time and the Displaced Rhythm Hypothesis

Peter Howell and his collaborators proposed a dramatic alternative to Node Structure Theory (e.g., Howell et al., 1983). Although his main interests lie in speech, and in particular the phenomenon of stuttering,

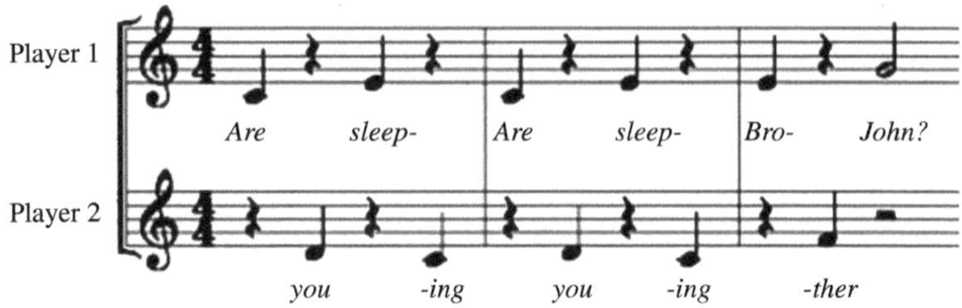

FIGURE 3.3 The phenomenon of hocketing, shown in musical notation. The two parts shown alternate to collectively produce the melody.

Howell maintains a keen interest in the psychology of music and as such his views were partly inspired by a musical phenomenon: *hocketing*. This musical practice originated in medieval times, and involves having one player alternate with the other in time. So player when player A finishes the first note, player B starts her first note, then player A starts note 2 after player B's first note, and so on. An example of this practice is shown in Fig. 3.3, using the familiar tune "Are you sleeping (brother John)?" The effect creates a kind of "hiccup" rhythm, which led to the name of the effect (*hocquet* for "hiccup" in French). Howell took note of the fact that hocketing is a bit like the effect of DAF. In the example of Fig. 3.3, for instance, the sounds of player B lag behind the actions of player A, a bit like DAF.

Two features of this analogy are critical. First, hocketing is incredibly hard to do because musicians are not used to coordinating rhythms in this way. Second, hocketing does not involve a manipulation of one's own auditory feedback! Howell suggests that the difficulty found in hocketing reflects the same constraints on perception and action that occur during DAF. He proposed, contrary to the prevailing view before 1983, that DAF effects have nothing to do with "feedback" in the literal sense. Perhaps DAF effects represent the disrupting effect of performing alongside any competing auditory rhythm.

In other words, DAF may be disruptive because speech, like music, is rhythmic. While executing the rhythm, an auditory pattern that conflicts, rhythmically, with your actions can cause interference. Any competing auditory pattern. This is not an outlandish idea. Try talking to someone while standing next to a person who is playing a tune on the piano that you *really* like—it can be difficult. Furthermore, DAF is particularly meddlesome because the competing pattern is the *same rhythm* you are trying to produce, but its onsets are *displaced in time*, relative to your own. Now imagine trying to talk to someone while your mischievous 10-year-old child stands next to you mimicking what you say a couple moments after you produce each word. (Again, this is not a very far-fetched scenario!)

This *Displaced Rhythm Hypothesis* has wide-ranging theoretical implications that go beyond DAF effects, and have been developed by Howell into a broader theory (EXPLAN) that I summarized in Chapter 1. For now, I will focus on the importance of its rhythmic basis. Howell made the strong falsifiable prediction that DAF disruption should be maximal when the *onset* times (i.e., sounded beginnings) for auditory feedback events coincide with the *offset* times for actions. Thus, if you are playing the trumpet, DAF will be most difficult to contend with if the sound of your trumpet blast occurs when you are reducing airflow at the end of each note. It just so happens that in speech, most syllables last about 200 ms. Because syllables constitute rhythmic units in speech, the syllable is the most plausible unit to stand as an "event" (see Chapter 1 for more discussion on what is meant by an "event").

Unfortunately the specific prediction made about the role of onsets and offsets may not do the trick. Evidence for this specific kind of an effect in speech is lacking, and Finney (1999) reported negative evidence for nonspeech sequences. Specifically, Finney analyzed the relationship between offset times in rhythmic tapping and which delay lengths caused the most disruption to performance. Offset timing did not predict which delays were disruptive. On reflection, even the original data from Howell et al. (1983) seem difficult to resolve with this notion. Experimental tasks in this paper included both speech and Morse key tapping. These tasks were disrupted by similar feedback delays, yet the durations of spoken syllables are often longer than those of key taps. Thus, according to the Displaced Rhythm Hypothesis, key tapping should have been disrupted by shorter delays than speech.

However, a broader point from Howell's work does hold up to scrutiny. Leaving aside for now the question of whether offset times are critical, an important point about this prediction is that it is based on *relative time* rather than the absolute time basis found in Node Structure Theory. A particularly compelling early demonstration of the role of relative time was published by Robinson (1972). He had participants produce lists of numbers that were displayed visually, one at a time. Participants were told to speak the number as soon as they saw it and so the rate at which the successive numbers appeared determined speaking rate. Robinson's measure of disruption came from the number of articulation errors participants made rather than produced speaking rate (which was not reported). According to Node Structure Theory, a delay of around 200 ms should maximally disrupt production regardless of speaking rate. However, a 180-ms delay (the closest delay to 200) only yielded maximal disruption for production rates in the vicinity of normal speaking rate (170–500 ms per syllable). For extremely fast rates (e.g., 100 ms per syllable) the most disruptive delay length was 90 ms and for extremely slow rates (e.g., 670 ms or 1 second per syllable), the most disruptive delay was 360 ms. Like Lee (1950), Robinson offered a mathematical summary of his results that offers a useful benchmark for subsequent research. Specifically, according to his data the critical delay may be predicted using the following equation:

$$D = \frac{1}{6} \times T + 160 \tag{3.2}$$

The units here are in milliseconds. D refers to a delay length that functions as a critical delay. T refers to the tempo, and functions similarly to T in Eq. (3.1), which had to do with total reading time. The other two terms come from fitting a linear function to Robinson's data but have broader implications. The slope, 1/6, can be interpreted as the *relative phase* at which the critical delay appears. Relative phase refers to where an event occurs within a recurring cycle; in this case the recurring cycle is established by the rate at which sequence events (for Robinson, syllables) are produced. However, the fact that this equation includes an intercept suggests that absolute time also plays a role. Interestingly, this intercept is fairly close to the time it takes to produce a syllable. So, for instance, if we assume that speakers in the studies listed in Fig. 3.1 produced speech at a rate of 200 ms per syllable, the critical delay predicted by Eq. (3.2) would be 193 ms, which is close to (though a slight underestimate of) MacKay's critical delay.

More recent research has further confirmed the importance of tempo in the effect of DAF for many tasks. This work has led to further theoretical developments and will be discussed later. For now, we will turn to a tentative resolution of the apparent paradox between these perspectives.

Resolving the Alternatives

At this point we find ourselves in one of those frustrating situations where one set of studies points to a given conclusion, and another set points to a different conclusion. Whereas Donald MacKay has published several results strongly suggesting absolute time (no effect of tempo on the critical delay), other results—Robinson (1972) being a salient example—support relative time. Is there a way to reconcile these differences? I propose that the differing results may reflect the temporal regularity of the produced sequence.

Recall MacKay's proviso that DAF effects depend on stimulus complexity (MacKay & Burke, 1972). In the context of speech, unfortunately, complexity comes with a cost. Syllables in speech are notoriously variable, particularly in stress-based languages like English (Patel & Daniele, 2003). Although repeating a single word, or two-syllable number sequences, may not represent the full complexity of speech, such manipulations do offer some control over the temporal complexity of the sequence.

Why is temporal complexity important for understanding the DAF effect? One reason has to do with the relationship between the content of auditory feedback and the content one produces. So far we have discussed DAF as if it were an entirely temporal phenomenon. However, under certain circumstances the delay may lead to a person hearing content from a *previously produced* event (e.g., a syllable) while attempting to produce the present event. Such circumstances are perhaps better considered as alterations of feedback content than of synchronization between perception and action.

And such a scenario is highly plausible! As a demonstration, I recorded myself speaking several sentences from MacKay's (1968) study that supported the notion of a critical delay based on absolute time. Fig. 3.4 illustrates the timing of my speech for a segment of one item. (The full item was "Although he should have known about it, pilots go to Mars.") My average interonset interval between syllables was just over 200 ms, as is typical for speech. The top row of text is aligned with the onsets of my speech, based on my segmenting of syllable onsets by hand in the software package Praat (Boersma & Weenink, 2013); the small dots show the exact timing of these onsets. The lower row of text, shown in gray, is aligned with where auditory feedback onsets should occur with a 200-ms delay, which is further highlighted by arrows. As can be seen, if I spoke with this timing while hearing DAF, some of the syllables I spoke (e.g., "known") would overlap with feedback from the previous syllable ("have"). Overall, 54% of the feedback onsets would occur after the onset of the next produced syllables.

Of course, we have noted that a common response to DAF is that the participant slows down. Thus, it is likely that the overlaps noted in

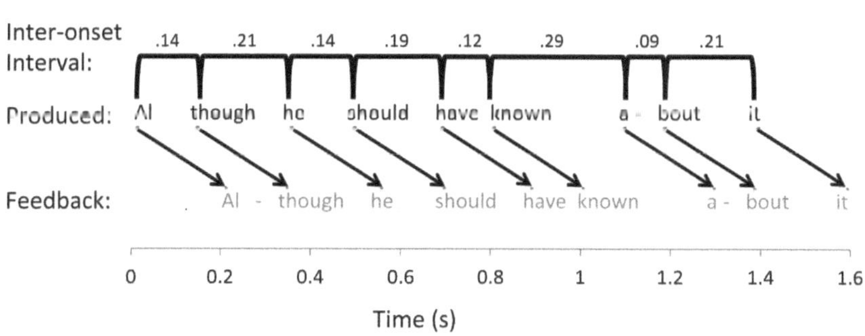

FIGURE 3.4 Timing of read speech by the author for a portion of the stimuli used in MacKay (1968), along with the approximate timing of syllables in auditory feedback given a 200-ms delay, shown in gray text.

Fig. 3.4 would not happen so frequently if I had been speaking with DAF. Nevertheless, the variable timing of syllables in this complex sentence could still lead to occasional overlaps of content between the current and previous syllables. Which leads to the second point: the variable timing of complex sentences may diminish the salience of rhythmic relationships between actions and auditory feedback. Although there is some reason to believe that speech production is guided by a meter-like framework (Brown, Pfordresher, & Chow, 2017), it is nevertheless true that speech rhythms are highly variable and the presence of anything akin to a "beat" is generally weaker, if at all present, than in music (Dalla Bella, Bialunska, & Sowinski, 2013; cf. Lidji, Palmer, Peretz, & Morningstar, 2011).

Ultimately, it seems to be the case that DAF responses reflect two processes, a duration-independent response similar to what MacKay proposed, along with a response to the relative phase of feedback onsets with respect to produced interonset intervals. When rate information lacks salience, the absolute time component dominates, whereas a salient rhythm promotes rhythmic binding between perception and action. In the next section I propose and evaluate a tentative model for how rhythmic binding between perception and action may work.

HIERARCHICAL PERCEPTION/ACTION COORDINATION: DELAYED AUDITORY FEEDBACK EFFECTS AS RELATIVE PHASE

I here discuss more recent research that suggests a different interpretation from the original studies. To a large degree, this newer research picks up where Howell's rhythmic displacement hypothesis left off. Whereas the older studies from that group focused primarily on the role of feedback content (or, more accurately, the absence of a role; see Chapter 4), recent research has focused to a greater degree on the kind of rhythmic relationships that determine what kind of an effect asynchronous feedback may have. These findings lead toward a model that forms part of the Hierarchical Perception/Action Coordination (H-PAC) hypothesis introduced in Chapter 1.

Disruption as Temporal Perturbation

Just like the term *feedback* suggests a certain kind of theoretical framework, so does the phrase *delayed feedback*. The theoretical implication behind this phrase is that we are working with a system that uses self-feedback for error correction, and that corrective feedback occurs later than it should. The effect of DAF, as noted earlier, may have nothing to do with the fact that feedback is delayed. Instead, as Howell and colleagues suggested, DAF effects may arise as a kind of rhythmic perturbation.

The EXPLAN model, discussed in Chapter 1, makes an important claim here. According to this model, any "...asynchronous inputs affect operation of the timekeeper in the execution system..." by way of increasing demands on that system (Howell, 2004b). The effects of such demands include slowing of timing and increased variability of the timekeeper (Howell & Sackin, 2002). Critically, a disruptive asynchrony may be interpreted as anticipating an onset, rather than a delay. Given that people respond to asynchronies as small as 4 ms when tapping to a metronome (Madison & Merker, 2004), and some effects (though modest) of DAF have been found for delays as short as 20 ms (Jack, Mehrabi, Stockman, & McPherson, 2018), a wide range of asynchronies could in principle influence performance.

Thus we have reason to believe that DAF effects arise because the onset times of perceptual events conflict with the planning and execution of actions, by virtue of the rhythmic relationships between perception and action. But how, when, and why do these conflicts occur? As noted before, the Displaced Rhythm Hypothesis does not seem to provide a full account, and EXPLAN is not specific on this matter. As such, I next turn to a proposal that is consistent in many respects to Howell's claims yet is based on a different kind of construct: Dynamical Systems Theory (DST). According to this proposal, the effect of DAF occurs because the timing of perceived onsets disrupts the phasing of an internal oscillatory timekeeper, based on unstable rhythmic coordination.

Dynamical Systems, Relative Phase, and Stable Coordination

DST is a branch of mathematics that has been adapted to various areas of cognitive science (Kelso, 1995; Large, 2010; Port & Van Gelder, 1995). In contrast to standard models in psychology, which rely on probabilistic processes that are usually independent of each other, the DST approach emphasizes deterministic systems that are highly interactive and often respond nonlinearly to variations of input variables. Many systems are based on rhythmic relationships, which may be stable (a state to which the system gravitates) or unstable (a state that is difficult to maintain). The temporal relationship between the planned timing of actions and disruptive DAF delays may therefore reflect unstable *rhythmic coordination.*

Fig. 3.5A illustrates how rhythmic movements may be represented as a periodic oscillation, as is common in DST. The cycle formed by

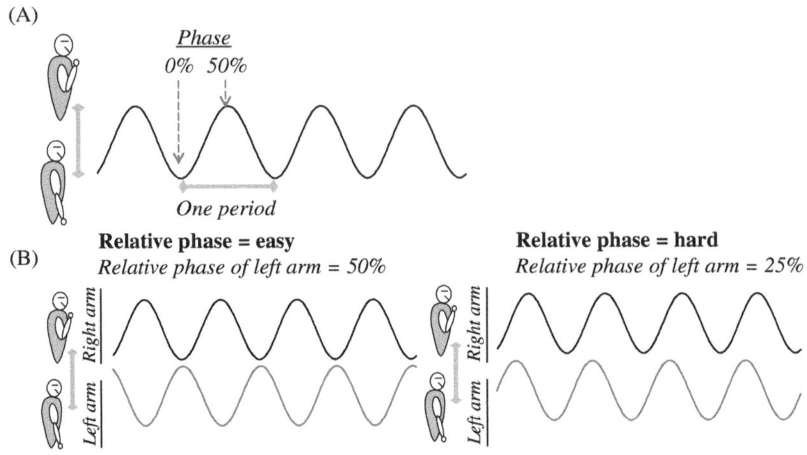

FIGURE 3.5 The phase and period of rhythmic arm movements in the sagittal plane (A). Examples of bimanual coordination where the relative phase of the left arm varies with respect to the right arm (B). Right arm trajectories are represented by the dark lines and left arm trajectories by the gray line, with peaks representing flexion and valleys representing extension in the sagittal plane.

swinging one's arm can be represented as a sine wave based on horizontal position of some reference point on the limb (e.g., the hand). At the start, the elbow joint is extended and the hand is at a low position. As the rhythmic cycle progresses, the arm moves up to the point at which the elbow is maximally flexed, which in the sine function is represented as a peak. The arm then progresses down to extension, at which point an entire cycle is finished. The time associated with a single cycle is its *period*, which may fluctuate over time but in Fig. 3.5A is simplified to be consistent. Within a period, we may also refer to different *phases*. Phases here are represented as proportions of the period (they may alternatively be represented in radial coordinates). In the figure, minimum points are considered to be zero phase because that is the starting point of the movement; peaks are at 50% phase because the movements here are regular and the points of flexion perfectly subdivided the period. If there were asymmetries in this movement, then the peaks may not be positioned at 50% phase. Of course, when the period varies, the absolute times associated with peaks may not be 50%.

There are many examples from research on motor control of rhythms that may be hard or easy to coordinate. Two contrasting examples are illustrated in Fig. 3.5B. Consider the coordination of two oscillating motor patterns, such as swinging your arms up and down at the elbow joint. These movements can vary with respect to their *period*, for example, you could swing your right arm at half the rate of your left arm. In Fig. 3.5B all arm movements have the same period. Alternatively, you can vary movements with respect to *phase* (starting point of the cycle). For instance you could start with your right arm extended (pointing to the floor) and your left arm flexed (hand up by the shoulder). This form of coordination is illustrated on the left side of Fig. 3.5B. As a result the left arm's position is always at the opposite point in the recurring cycle relative to the right arm's position, and we can refer to the *relative phase* of the left arm as being offset by 50% of the right arm's phase. This form of coordination is easy to maintain. People routinely engage in this kind of movement when walking or running. Studies in which participants are asked to coordinate rhythmic movements of limbs suggest that this kind of relative phase is easy to acquire and maintain, and also easy when two people have to coordinate rhythms with each other (cf. Demos, Chaffin, Begosh, Daniels, & Marsh, 2012; Schmidt, Carello, & Turvey, 1990). In music, this sort of coordination is akin to a reggae guitarist strumming chords on the "offbeat" relative to the rest of the rhythm section. By contrast, the kind of rhythmic coordination on the right of Fig. 3.5B is much harder to maintain, in which the phase of one arm is shifted by 25% of the other period.

It is important to note that one does not need to adopt a DST perspective to address rhythmic coordination. To some degree, patterns of coordination may simply reflect stochastic noise in the signal sent from the brain to one's limbs, as well as variability in one's internal timekeeper (Wing & Kristofferson, 1973a, 1973b). What the DST framework adds is the notion that certain forms of coordination are inherently stable, and thus may function as an *attractor state*. When the system is in one of these states it tends to stay there, because it is attracted to that state, and likewise if one nudges the system out of that state (e.g., someone bumps your arm), you will drift back to that state without knowing it. The kind of coordination represented on the left side of Fig. 3.5B would be considered stable, though not as stable as coordinating movements perfectly in phase. In contrast to stable states, coordinating in an unstable way (e.g., right of Fig. 3.5B) is difficult to maintain and individuals usually drift toward a more stable state if their coordination is perturbed in any way.

Auditory Feedback and Rhythmic Coordination

What does all of this have to do with effects of asynchronous feedback? Following the lead of Finney and Warren (2002), I propose that auditory feedback should be conceptualized as a signal that interacts with an internally generated period (an oscillation) that the performer uses to time planned actions. The notation that action planning may be based on such internal rhythms comes from the seminal theory of dynamic attention developed by Mari Riess Jones and colleagues (Jones, 1976; Jones & Boltz, 1989; Large & Jones, 1999). According to this view, effects of DAF do not occur because the performer has to wait for confirmation concerning the realization of a plan, but instead reflect the difficulty of coordinating two asynchronous rhythmic patterns. One pattern is based on movement; the other is based on the timing of auditory events. Moreover, patterns of disruption from DAF may reflect the presence of stable or unstable states of coordination between perception and action.

Recall that patterns of stability are not limited to situations in which individuals coordinate different rhythms that they themselves create (as in different arms)—similar patterns are also found when two people coordinate together. In such cases, coordination exists between perception and action. I propose that a similar process goes on when auditory feedback is asynchronous from actions. Importantly, as suggested by Howell (2004b), this view does not assign any importance to "feedback" per se. Indeed, if this view is correct then similar effects should arise even when asynchronous feedback does not sound like the kind of auditory stimuli you expected to produce, which is the case, as we will discuss further in Chapter 4.

According to the H-PAC model, effects of altered auditory feedback come from our attempt to coordinate perception and action with varying degrees of success. According to this view, auditory feedback serves a very minor role in the timing and sequencing of actions, particularly for well-learned sequences (cf. Chapter 2). Instead, feedback is treated as a pattern of sounds that one coordinates in time with the timing of planned actions, and is thus not substantially different from sounds that may be created by accompanying musicians during a performance. Furthermore, patterns of disruption when sounds are asynchronous with actions may be based on properties of rhythmic coordination that may be seen in other behaviors, possibly not limited to musical behaviors.

This hypothesis is based on the idea that the asynchrony disrupts the phasing of an internal oscillator (the timekeeper). This leads to a lengthening of the IOI surrounding the asynchrony that is proportional to the magnitude of the asynchrony. Importantly, phase has no "memory," so the lengthening of IOIs is stationary (does not build over time) and temporary (as mentioned above). Computationally, we may represent the hypothesis as follows (cf. Large & Jones, 1999; McAuley, 1995):

$$\mathrm{IOI}_{n+1} = p_0 + p_0\eta\left[\frac{1}{2\pi}\sin 2\pi(\phi_n - \phi_T)\right] \tag{3.3}$$

where p_0 indicates a fixed, that is, nonadaptive, oscillator period (a simplifying assumption I will take on later). The parameter ϕ_n relates to the timing of feedback onsets relative to the period of the current IOI. Timing is thus based on the *relative phase* of feedback onsets, which in the function range from −.5 half a cycle early to 0 (in sync with the oscillator) to +.5 (half a cycle late). The parameter ϕ_T is a threshold establishing the minimum amount of delay (i.e., the minimum relative phase) that can have an effect. The rest of the terms inside the square brackets map this discrepancy of relative phase on a position within a recurring cycle, represented as a circle (hence the use of pi and the sine transform). The parameter η models how strongly the oscillator responds to differences of relative phase. In a sense, this is just a fancier version of Robinson's model shown in Eq. (3.2). That model also proposed a threshold delay amount (the y-intercept), and had a slope term that reflected relative phase. The main (and important) difference is that this model is nonlinear, per the effects of asynchronous feedback seen in many studies.

How does this framework simulate effects of asynchronous feedback? Fig. 3.6 illustrates how a single asynchrony may influence timing according to the model. [Note that the bracket in Eq. (3.1) is expressed as $f(\phi_n)$ to save space.] Onset times associated with three actions (e.g., finger taps) are shown, which form two IOIs, along with auditory feedback timing for the first two actions. When there is no feedback

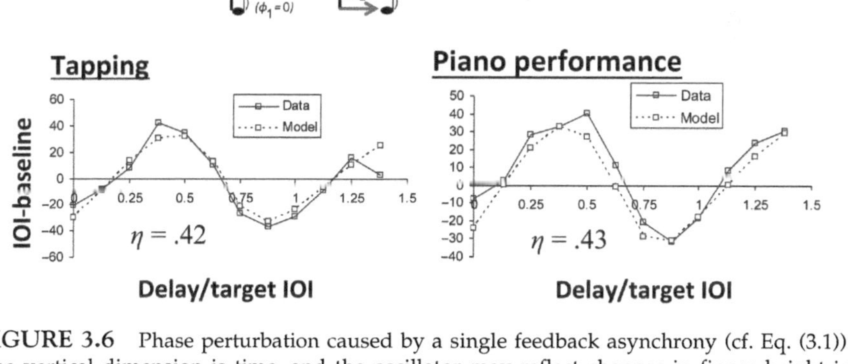

FIGURE 3.6 Phase perturbation caused by a single feedback asynchrony (cf. Eq. (3.1)). The vertical dimension is time, and the oscillator may reflect changes in finger height in space. The timing of auditory feedback is indicated by the note heads, and the effect of a delayed onset for the second IOI is shown by the addition of the gray vertical line to the base period of the oscillator. See the text for further details. Lower panels show fits of this model to performance data (Pfordresher, 2007).

asynchrony (tap #1), the resulting IOI is equivalent to the base period of the oscillator (p_o). However, when there is an asynchrony (tap #2) that causes feedback onsets to fall within the first half of the IOI, $f(\phi_n)$ is positive. This leads to a delay in the subsequent tap (tap #3), which lengthens the IOI. Fig. 3.6(top) shows an extreme case in which coupling is maximal. This causes the expansion of the second IOI to be fully equivalent to the phase shift brought about by feedback asynchrony. However, the base period of the oscillator (p_0) does not change, so any future effects of asynchronous feedback will occur independent of the effect found for the present IOI. A different effect can be found for asynchronous feedback that occurs in the second half of the IOI. These asynchronies lead to a temporary shortening of the oscillator period because $f(\phi_n)$ in these cases is negative. The lower part of Fig. 3.6 shows fits to data from tapping (left) and piano (right) performance tasks, where coupling is less than 100%. *Y*-axes in each case show difference in mean produced IOIs (the tempo) during that condition to a baseline measure of production made earlier in the session, in milliseconds. In the next few sections I consider in more detail how well the assumptions of this very simple model cohere with available data.

EVALUATING HIERARCHICAL PERCEPTION/ACTION COORDINATION

I here review research that addresses four predictions that arise from the proposal articulated in the previous section. This research is dominated by piano performance and rhythmic tapping tasks. Although a focus on these tasks is appropriate for this book, it also reflects an inherent advantage for music. As noted before, most speech has highly complex temporal structure (e.g., Fig. 3.4). In addition, it is hard to identify precisely where syllable onsets occur, either in production or perception (Scott, 1998). By contrast, playing the piano and tapping on the table involve percussive movements that lead to sounds with a clearly defined onset in the amplitude envelope (cf. Vallet, Shore, & Schutz, 2014).

Prediction 1: Disruption Should Be Based on Relative Time

The question of whether the appropriate parameter to account for disruptive effects of DAF is based on relative or absolute time was discussed earlier. H-PAC sides definitively with the relative timing view because relative phase (the critical measure) is a dimensionless proportion. I discussed earlier research on this topic. However, the ability to control and manipulate timing precisely has increased dramatically since that time and so it makes sense to reevaluate the role of relative and absolute time. As we will see, the relative time prediction fares well, although there does appear to be some constraining effect of absolute time as well.

Finney and Warren (2002) reported strong support for the relative time hypothesis in a paper that arose from the first author's dissertation (Finney, 1999). Their work adopted the logic of Robinson's (1972) aforementioned study, which demonstrated that the "critical delay" may shift to longer delays when production rate slows down. However, Finney and Warren incorporated a task with greater temporal control that led to a fascinating new discovery. Participants tapped a regular but nonisochronous rhythm. The rhythmic involved an alternating group of 4 IOIs (interonset intervals) with a group of 2. In a musical context, the rhythm would be counted like this: "One, two, three, four, (rest) two, three (rest), one, two, three, four..." and so on.

The use of this rhythmic pattern allowed the authors to test a unique hypothesis: What happens when the delay is equal to the base IOI (i.e., the IOI occurring between onsets that are not separated by a rest) and occurs at a point where no movement onset is planned (the "rests")? There is an interesting paradox here. On the one hand, DST theories would suggest that this kind of phasing leads to stable coordination. With respect to the earlier example of arm movements, a delay equal to one full period would be comparable to initiating left arm swings after one full period of the right arm. This is not hard to do. Thus, one might predict that delays equal to the base IOI would not disrupt performance. On the other hand, feedback onsets in this scenario are occasionally synchronized with positions where no tap is intended. According to the ideomotor view discussed in previous chapters, the event perceived at this time should be associated with a movement that the performer is supposed to suppress. To top it off, a DST-inspired theory of auditory attention, called the *Dynamic Attending Theory*, would predict that one's attention to incoming events is maximal at times matching the base period of a sequence (Jones & Boltz, 1989; Large & Jones, 1999). According to this view, attention is most strongly focused at times where we anticipate that an event may occur.

Finney and Warren found that delays equal to the base IOI did cause disruption, although the sort of behavior they observed was not similar to "disruptive" patterns reported in other reports. Instead of causing participants to slow down, the critical delay caused participants to insert additional (unintended) taps at points where the pauses were. Interestingly, the taps did not generally follow after the delayed onset and thus were not obviously triggered in a manner consistent with Node Structure Theory (Finney, unpublished manuscript;

cf. MacKay, 1987). Also, participants would tend to pause after these inserted taps, suggesting a general phase shift of the rhythmic pattern. What seems to happen here is that the participants synchronize with the base IOI, but are also sensitive to the higher-order rhythmic pattern. When that pattern is shifted, the pattern of tapping reflects confusion about the starting point of the pattern, leading to inserted taps.

Beyond this, the most important contribution of Finney and Warren's work is that it showed a clear shift of the "critical delay" for DAF. Participants tapped this rhythm at two tempos. The critical delay hypothesis would predict that tapping tempo should not influence the effect of different delay conditions. However, the aforementioned insertions were found when delays were equal to the base IOI, and thus the effect of delay did vary with tempo. Moreover, disruption was not maximal at delays of a fixed absolute magnitude, such as 200 ms. Thus, their study shows a clear role for relative timing, consistent with H-PAC.

By analyzing insertion errors, Finney and Warren cleverly circumvented a practical issue that arises in standard effects of the manipulation, which may confound the relative phase of auditory feedback. Consider the fact that DAF typically causes participants to *slow down* when it disrupts timing. Because of this it can be impossible to account for DAF effects based on relative phase with optimal accuracy and precision. For instance, assume that you predict that disruption may be high when the relative phase of auditory feedback is based on 25% of the period between piano keypresses, similar to the right side of Fig. 3.5B. If you have the participant play at a tempo with a beat period of 600 ms, a delay of 150 ms would be positioned at this relative phase. However, as we know, once the participant hears these delays he or she will slow down. If this slowing is substantial, it may have a noticeable effect on the relative phase of auditory feedback.

How can one deal effectively with such a complication? Fortunately, the use of MIDI technology in the software program FTAP, created by Finney (2001a), allowed a solution to this problem. FTAP is a Linux-based program that reads a sequence of messages from any MIDI input device (e.g., an electronic keyboard), records and processes each input message (including possibly altering the content, e.g., pitch value), calculates when each corresponding output message should be sent (e.g., immediately, or with a delay), and then sends the message at the appropriate time to an external MIDI tone generator. The entire loop can be completed within about 1 ms given the appropriate equipment (Finney, 2001b, 2016). Importantly, one feature of FTAP allows the length of the delay to be based on a proportion of the local tempo, which can be estimated using the average of N recent interonset intervals, disregarding outliers.

A particularly critical test of the role of relative time was reported by Pfordresher and Benitez (2007), who had participants produce sequences (tapping or melodies on a keyboard) at three different tempos. When delays were based on relative phase (referred to as *adjustable delays* in the paper), their effects were not altered by production tempo, whereas delays of the traditional sort—based on absolute time—had effects that did covary with tempo.

That being said, Pfordresher and Benitez also found that absolute time had a generally constraining effect on performance. Although relative time played arguably the strongest role in their data, it was also the case that disruption was limited to delays that were shorter than 400 ms. Although this result is not consistent with absolute time views like that of Node Structure Theory (MacKay, 1987), it does suggest that some kind of absolute limits exist on the tendency for asynchronous event onsets to disrupt performance. For instance, an unpublished study from my laboratory revealed effects of asynchronous auditory feedback among trained pianists when performing at rates of 500 or 700 ms per IOI (i.e., 120 and 85.7 bpm), but no effect of asynchrony (and generally high timing variability) for faster or slower rates (Pfordresher & Flaherty, unpublished data).

Prediction 2: Effects Should Emerge and Dissipate Rapidly

If effects of asynchronous feedback are due to phase perturbations of one's internal timekeeper, then their effects should be highly contingent on the presence of that perturbation. That is, effects should appear almost instantaneously once the asynchrony is presented, and likewise effects should disappear almost instantaneously once the asynchrony is removed.

At this point, it is important to note that the DST approach offers two hypotheses about the basis of asynchronous feedback effects. The timing of oscillations are defined by both phase and period. The fact that asynchronous feedback leads to persistent slowing of IOIs could be interpreted as a slowing of the inner period of the oscillator based on the phase perturbation. One can model the effects of changing the period simply by having the period of the oscillator respond to asynchronies rather than remain fixed. Thus a changeable period would replace p_0 in Eq. (3.3). There are two important implications of the alternative model. First, the effect of asynchronous feedback should persist after the asynchrony is removed. So if a pianist slows down while hearing DAF, they will continue to play slowly if feedback timing returns to normal. Second, the effect of asynchronous feedback may accumulate. This is particularly true of adjustable delays that retain a consistent relative phase each time they occur. In such cases, every onset will be followed by an asynchrony of roughly consistent phase, causing the internal period to slow down or speed up every time such an asynchrony is experienced.

Pfordresher and Kulpa (2011) tested these hypotheses. They presented feedback asynchronies set to be 33% of IOIs during rhythmic tapping or performances of piano melodies. Most important, trials were divided into three phases. Participants would experience normal feedback during the first phase, then would experience asynchronies during the second phase, and then finally the asynchrony would be removed during the third phase. The length of phase 2 varied to evaluate whether asynchrony effects changed over time. The results were clear: asynchrony effects emerged immediately and then disappeared rapidly (i.e., within 1–2 onsets) after the asynchronies ceased, thus supporting the phase perturbation model of Eq. (3.1).

This result offers strong support for prediction 2. Results from other papers suggest that alterations to period may occur, although these results are not reliable enough yet to establish precisely the effect of auditory feedback on phase versus period. One possibility is that DAF affects the period of speech more strongly than the period of music production. For instance, Black (1955) reported persistence of the DAF effect on speech for a little over 2 minutes, although close evaluation of the data suggests that this effect was highly variable across participants and over time. Several papers have addressed longer-term adaptation of speakers with DAF, although none approximates the timescale of expert pipe organists. Three studies showed decreased effects of DAF during multiple extended sessions in which participants read words or prose passages. The advantage to repeated exposure found in these studies was limited, however. Tiffany and Hanley (1956) found a reduction in "fluency breaks" (stuttering-like responses) during prose reading, whereas DAF continued to slow overall production rate; another study found stabilization of reading rate with DAF during prose reading (Goldiamond, Atkinson, & Bilger, 1962). Katz and Lackner (1977) replicated improvements to prose reading, but found that this learning effect did not extend to repeating word lists. Taken together these results suggest that alleviation from DAF may come about due to enhanced sequential planning, given that sentences involve higher-level syntactic organization (in contrast to word lists) and fluency breaks suggest breakdowns in planning. At the same time, no studies seem to exhibit prolonged slowing effects of DAF, in keeping with the relative phase hypothesis described above. In particular, Katz and Lackner (1977) found that participants sped up after hearing DAF, much like the results of Pfordresher and Kulpa (2011).

Within music, one real-world example suggests that long-term training may involve sensorimotor adaptation to DAF. Consider an organist who performs in a large cathedral. Mechanical pipe organs often have a considerable lag time between the musician's keypress and the sounding of the pips, which can be positioned at considerable distances from each other leading to delays that can be as long as 150 ms (Pollard, 1968; Wolley, 2006). These musicians are often of high caliber, and do not perform in ways that resemble typical effects of asynchronous feedback.

To date, evidence suggests that DAF affects the phasing of rhythm production, with possible effects on period with sustained exposure. These long-term effects are borne out in a reduced sensitivity to delays, which could be modeled by changes to ϕ_T in Eq. (3.3).

Prediction 3: Release From Disruption Should Occur if Asynchronous Feedback Coincides With a Stable Coordination Mode

As mentioned in my discussion of Fig. 3.5, not all asynchronies are problematic for coordination. In fact, a major finding from research on bimanual coordination is that *antiphase coordination* is often stable. As shown on the left side of Fig. 3.5B antiphase coordination occurs when the starting point of one movement cycle coincides with the midpoint of another movement cycle. Although there are many studies verifying this generalization, a particularly famous demonstration was reported by Haken, Kelso, and Bunz (1985), and involved having participants wiggle the index fingers of their right and left hands, back and forth to the left and right (as one does when pointing at people and chanting, "Eenie, meenie, miney moe..."). With respect to the muscles involved, in-phase wiggling occurs when both fingers point to the center at the same time, whereas parallel movement is considered antiphase coordination. Antiphase coordination can be as stable as in-phase coordination, but only at slow rates. As people speed up the rate of their alternating finger wiggles, antiphase coordination gets harder and people end up switching over to in-phase (i.e., fully synchronized) coordination.

When I first began studying effects of DAF with Caroline Palmer, my primary interest was on determining whether stable coordination modes existed that alleviate the effect of DAF. In other words, if you can arrange the timing of feedback onsets to form antiphase coordination, would disruption diminish, or even disappear completely? The answer was rather disappointing. Though this sort of effect has emerged in some experimental conditions, it is unreliable. For instance,

Pfordresher and Palmer (2002) found that a 250-ms delay led to relatively less disruption than other delays when the rate of performance was 500 ms per IOI, where disruption was defined as an increase in timing variability in keyboard performance. However, this effect was not replicated in a second condition in which performers produced sequences at a rate of 600 ms per IOI (in which a drop in disruption should be found for a delay of 300 ms). In a similar follow-up study using adjustable delays (described earlier), Pfordresher (2003a) found no evidence for alleviated disruption with delays timed to be 50% of IOIs.

Part of the unreliability of this effect is likely due to the complexity of movements underlying piano performance. Although the piano sequences in the two studies described in the previous paragraph both involved isochronous timing of IOIs, the finger movements leading to those IOIs was sequential and involved the kind of complex spatiotemporal patterns described in the previous section. In other words, coordination may be based to a great degree on periodicities in movement, rather than periodicities in onset times.

If this is the case, then there may be more evidence for stable coordination modes in rhythmic tapping. And there is. As detailed below, Pfordresher and Dalla Bella (2011) found differing effects of asynchronous feedback based on whether it occurred before or after the midpoint of movement during tapping. How about delays coinciding with the midpoint of movement, when the finger reaches its maximum height and velocity switches from upward to downward? As it turns out, the effect of asynchrony disappeared for these kinds of asynchronies, which is consistent with predictions of DST. But taken together, evidence for prediction 3 is relatively weak.

Prediction 4: Disruption Should Be Associated With Movement Patterns

As described above, DST theories of coordination are based on synergies that exist across groups of muscles. Although there have been applications of DST ideas to cognition that are not based on motor control (e.g., dynamic attending theory), when it comes to an area like music performance it stands to reason that dynamic coordination between sound patterns and motor planning may be rooted in how tone onsets interact with movement patterns.

A critical test of this view was conducted by Pfordresher and Dalla Bella (2011). The authors used motion capture during a rhythmic tapping task to measure the position and velocity of the participant's finger at the time when an asynchronous auditory feedback onset would occur. They incorporated the adjustable delays described earlier, so that this relationship would remain roughly stable through the course of an experimental trial. It is worth noting that the task used in Pfordresher and Dalla Bella was simpler than the other tasks described so far, and could only be considered "musical" in the broadest possible sense. Participants simply listened to a regular metronome click, tapped in synchrony with this click, and then attempted to maintain the same regular pattern after the metronome stopped—possibly while having auditory feedback delayed. Thus, the temporal pattern was *isochronous*, consisting of a single unchanging time interval, and there were no sequential patterns of movement to plan. Though simple, this task has been used often in studies of motor timing due to the high degree of control it offers. Also, believe it or not, people are far from perfect at this task. Timing is variable, even among the most well-trained musicians, and sometimes people drift in their tempo, which is what the authors measured.

As we have come to expect, participants in this study reliably slowed down for asynchronous feedback. However, not all delays led to this effect. In fact, very long delays—delays so long that they approached the time of the participant's subsequent tap, significantly *sped up* the rate of tapping. A similar effect arose in the data presented in the lower part of Fig. 3.6. Thus, this study found that asynchronous feedback could lead to two opposite responses in participants—either speeding up or slowing down. This finding on its own renders problematic the typical characterization of feedback alteration effects as being "disruptive." Usually a disruptive effect goes in a single direction: children fail to stay quiet, a circuit board shorts, a supply chain breaks down. Here we have evidence that asynchronous auditory feedback *perturbs* timing in a certain direction. And the basis for this perturbation seems to be based on the relationship between feedback timing and movement.

Imagine yourself tapping on a table. Similar to the rhythmic arm movements shown in Fig. 3.5, this movement involves a regular oscillating pattern that may resemble a sine curve (although the peak may occur somewhat late in the period, leading to a slightly asymmetry; Balasubramaniam, Wing, & Daffertshofer, 2004). Now consider the relationship between this movement and the onset of delayed auditory feedback. If the relative phase of this onset occurs at the halfway point in your movement cycle, it would sound when your finger reaches its peak position. This is analogous to the antiphase coordination pattern between arm movements shown on the left side of Fig. 3.5A. Delays shorter than this would be associated with upward movement of the finger, and delays longer than this would be associated with downward movement of the finger. When delays were shorter than the midpoint of the finger's recurring cycle—that is, when delays sounded as the finger was on its way up—tapping rate slowed down. This effect is what one usually finds in earlier DAF studies. However, when delays were longer than this midpoint—that is, when delays sounded as the finger was on its way down—tapping rate sped up. The opposite of what DAF usually does.

Thus, asynchronous feedback's effect may be based in part on the associations one has between the onsets of sound and the velocity of movement. For tapping, sound onsets are associated with downward motion. Thus, if sound onsets occur as the finger moves downward, auditory feedback acts as a force that propels the movement forward given its association with the forthcoming goal. However, if the feedback onset occurs as the finger moves upward, the association between sound and movement conflicts. The sound onsets thus exert a kind of breaking force on movement, leading to slowing.

Do these results from simple repetitive finger tapping generalize to more complex movements used in music performance? No motion capture data yet exists; however, results from the aforementioned study by Pfordresher and Benitez (2007) suggest that effects of asynchronous feedback may vary across tapping and keyboard performance in a way that reflects the different movement patterns associated with each task. Earlier, I highlighted trials from that study that involved performances of melodies on a keyboard. However, Pfordresher and Benitez also examined repetitive isochronous tapping. Movements in keyboard performance differ from the patterns of movement seen in finger tapping. During piano performance, movements are sequential and fingers are typically moved from one key to another during the performance. Research from Caroline Palmer's laboratory at McGill University has documented in great detail the finger movement patterns characteristic of keyboard performance (Dalla Bella & Palmer, 2012; Goebl & Palmer, 2008, 2009; Loehr & Palmer, 2009). To prepare for such variable movement patterns, pianists typically move fingers a few notes in advance of when they will be used. The finger than hovers above the key until very shortly before the keypress occurs. A few milliseconds before the intended onset, the finger plunges down rapidly toward the keybed, with the velocity being influenced by the intended loudness of the note as well as performance tempo (Dalla Bella & Palmer, 2012). The pattern of results shown in the lower half of Fig. 3.6 resembles this pattern (though the results are much subtler), in that a stronger effect is found for shorter delays in piano performance than for simple tapping.

THE NEURAL BASES OF TEMPORAL BINDING BETWEEN PERCEPTION AND ACTION

In contrast to the neuroscientific research on the formation of perception/action associations, discussed in Chapter 2, very few studies have addressed how the brain responds to asynchronous auditory feedback. In fact, only one neuroimaging study to my knowledge addresses this effect in music performance (Pfordresher, Mantell, Brown, Zivadinov, & Cox, 2014), and three have addressed the effect of DAF on speech (Hashimoto & Sakai, 2003; Takaso, Eisner, Wise, & Scott, 2010; Watkins, Patel, Davis, & Howell, 2005). In all four of these studies participants produced sequences (melodies or text) under various auditory feedback conditions, including conditions with a delay of auditory feedback and normal auditory feedback. All four studies likewise published analyses in which brain activity during production with a delay was contrasted with production during normal feedback. Note that in these contrasts, both conditions include motor production and auditory perception.

Fig. 3.7 illustrates two brain areas that appear consistently across these studies. It is useful to compare this figure with Fig. 2.2, which summarized brain areas that responded to the formation of motor associations with perceived pitches as a result of training. Again we find activation in the inferior frontal gyrus (Broca's area), although results for the effect of asynchronous feedback are more variable than was the case for studies discussed in Chapter 2. The most critical study here is Pfordresher and colleagues, who used musical materials. Activations in this area were found bilaterally, but were stronger on the right than the left side (Fig. 3 of their paper only shows right-side activation). The presence of this area in both cases suggests that asynchronous feedback may be due to the fact that asynchronies reactivate sensorimotor associations more strongly than listening to normal feedback, similar to MacKay's (1987) claim about asynchronies occurring during a content node's hyperexcitable phase.

More consistent activation was found in the superior temporal gyrus, a bump that runs lengthwise along the temporal lobe, which plays an important role in many higher-level auditory functions. We did not see activations in this area for the studies summarized in Chapter 2, even though those studies—like the present ones—contrasted two experimental conditions that both involved auditory perception. For some reason, the auditory system responds more strongly to delayed auditory feedback than regular auditory feedback during performance. Hashimoto and Sakai (2003) suggested that this activity relates to heightened awareness of and attention to delayed feedback, an account that is

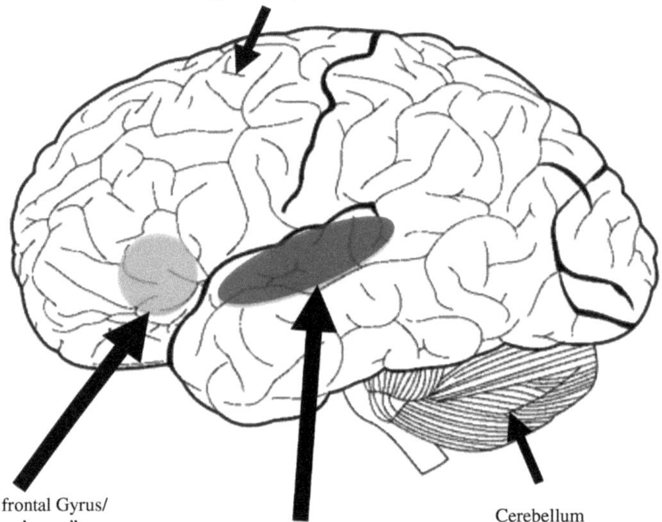

FIGURE 3.7 Approximate location of brain areas exhibiting larger responses during production with asynchronous auditory feedback than during production with normal feedback in music performance (Pfordresher et al., 2014) as well as in speech. Activation in the darker oval was corroborated in three different studies of speech (Hashimoto & Sakai, 2003; Takaso et al., 2010; Watkins et al., 2005), whereas activity in the lighter oval was only found in two speech studies (Hashimoto & Sakai, 2003; Watkins et al., 2005). Superior temporal gyrus activity was bilateral in all studies, whereas inferior frontal gyrus activity varied in lateralization. Small arrows are used for areas specific to Pfordresher et al. (2014) and, in the case of the cerebellum, only one speech study (Watkins et al., 2005).

consistent with the idea that the delay leads to a release from the auditory system's tendency to suppress the perception of self-feedback.

What about neural centers for timing? Although the neural representation of time is not a settled issue, two brain areas seem to be the top contenders for the brain's "internal clock": the cerebellum and the basal ganglia. Neither of these areas appeared consistently across the studies reviewed here, although greater activation in the cerebellum was found during production with DAF in two studies (Pfordresher et al., 2015; Watkins et al., 2005). In my study, we did actually find some activation in the basal ganglia, but not at a high enough level to be considered statistically significant. Thus, although asynchronous feedback disrupts timing, it might not lead to greater activity in timekeeping centers. Instead, the coordination of timing across perception and action, mediated by the inferior frontal gyrus, may be the primary effect.

CONCLUDING REMARKS

In this chapter, I summarized the research on how asynchronous auditory feedback influences music performance and the theoretical implications of these effects. In so doing, I introduced a detailed account of how these asynchronies may reflect the functioning of an integrated timekeeper used for perception and action, which is one component of the H-PAC model introduced in this book. Studies of feedback asynchrony have been going on longer than any other kind of effect I discuss in this book, and as a result this was a long chapter. Even so, as we have seen, there are questions that remain for future research, particularly concerning the balance between absolute and relative timing. In the next chapter we will step up one level in the planning hierarchy to consider how perception and action are integrated with respect to the serial ordering of content.

CHAPTER

4

Binding Planned Actions to Their Consequences

In 2014 the British composer Jonathan Pitkin created a traveling art installation called the "misbehaving piano." Using a computer program called Max-MSP, Pitkin rewired the associations between keys on the piano and the pitches that would result. A variety of scenarios were modeled in this way. In one scenario, the keyboard would generate the pitch of an incorrect note at an unexpected position in the performance. This pitch output was generated independently of whatever was performed. Thus if a visitor to the installation performed with complete accuracy, at one point auditory feedback would indicate that, contrary to the facts, he or she had produced a "clunker." In another, more pernicious scenario, called "lag," the pitch heard at each keypress would match whatever pitch should result from the immediately preceding keypress. So the output pitch would always be one note behind what the participant produced. An entertaining YouTube video of the traveling exhibition shows how participants responded to these conditions—generally with great amusement and often befuddlement.[1]

By coincidence, the composer had rediscovered many altered feedback conditions that had been used in psychology experiments, although participants in these experiments typically do not respond with the same kind of amusement one sees in the videos for the misbehaving piano. A common theme to these manipulations is that they involve alterations to the *pitch content* of auditory feedback while retaining synchronization between actions and sound. Thus, these kinds of alterations are conceptually distinct from the manipulations of feedback synchrony discussed in the previous chapter. Whether these different

[1]Some performing musicians alter the mapping of pitch on their own instrument, referred to as "voluntary self-sabotage" by De Souza (2017), because of the cognitive stimulation that such alterations bring about.

Sound and Action in Music Performance
DOI: https://doi.org/10.1016/B978-0-12-809196-8.00004-8

manipulations influence different mechanisms for relating perception and action in music constitutes the big question at hand.

As in other chapters, the research summarized here will be dominated by studies of piano performance. However, a sizeable literature also exists concerning the role of pitch content in vocal motor control, focusing primarily on fine-grained pitch control. Because many of these studies concern associations between spoken and sung pitch, I will discuss the vocal pitch feedback literature in Chapter 7.

THE MEANING AND SIGNIFICANCE OF "CONTENT"

It is worth considering why we may want to distinguish the *content* of auditory feedback from its synchronization with actions. With respect to effects of altering feedback, for instance, it is plausible that performers may treat any alteration as something gone amiss, and will simply "mess up" in response. In such a scenario, effects of altering synchrony or feedback content would be indistinguishable.

First, let me explain why I am introducing a more general term like *content* as opposed to simply referring to pitch. As in Node Structure Theory (see Chapter 1), *content* here refers to the category associated with an event in auditory feedback, the "what" instead of the "when." Although there is evidence of categorical perception for rhythms (Clarke, 1987), there is no sense in which synchronization of an action with its perceptual consequences constitutes a category. By contrast, pitch constitutes the most salient category structure in music, and thus provides an excellent starting point to understand the role of feedback content in relating perception and action. Musical pitches have category boundaries based on intonation (in-tune vs out-of-tune), and tonal structures provide a rich hierarchical framework to make distinctions among different category members (Krumhansl, 1990). Although there are other features that categorize musical events—timbre being the most salient example—pitch is the most well-understood musical dimension and happens to be what studies on the role of auditory feedback have explored. Moreover, in talking about content rather than just pitch, we are able to draw comparisons between music and language, given that speech sounds in English are categorized primarily by spectral rather than pitch content (see Chapter 7 for more discussion of speech).

Now let us consider why feedback content merits separate discussion from synchronization. Synchronization, as discussed in the previous chapter, operates on fairly rudimentary principles that may be governed independently from the sequential structure that variations in pitch give to a melody. In fact, a great deal of research in psychology has shown that melody and rhythm (i.e., the temporal pattern of onsets) contribute separately to melody recognition, although studies differ with respect to whether this contribution is fully independent or interactive (see discussions in Jones, 1987; Krumhansl, 2000; Palmer & Krumhansl, 1987; Prince, Thompson, & Schmuckler, 2009).

Another reason to separate content from timing in perception/action associations has to do with the nature of the sensorimotor associations involved. In Chapter 3 I argued that temporal associations between perception and action may be guided by an internal timekeeper that responded jointly to planned and perceived event onsets—an oscillator. It is hard to see exactly how such a simple framework could be used to predict how we relate the anticipated outcomes of our actions to what we ultimately perceive. Ultimately, associating perception and action based on content may involve higher-level cognitive processes. Binding by content requires that we have some kind of learned association that causes us to expect that a certain perceptual event should result from the execution of a planned action. Thus, on a piano the goal of pressing a single key must be associated with the kind of sound that should result, as we discussed in Chapter 2 in the context of ideomotor planning.

When it comes to producing a melody, the issue becomes even more complex. Melodies are planned *hierarchically*, and so associations may be based on relating a sequence of planned actions to a sequence of auditory events. As discussed in Chapter 1, melodies are also planned *incrementally* (like speech), such that a performer assembles the plan for one sequence of actions (e.g., a musical phrase) during the execution of a preceding sequence of actions. Thus, it is important to consider effects of auditory feedback not just for the planning and execution of isolated events (notes or chords), but also for the production of full melodies. Theoretically, it is important to consider how well constructs like ideomotor planning can account for effects that relate to entire sequences. Nevertheless, it makes sense to start the discussion at the simplest level, and so this chapter begins by reviewing evidence from studies that focus on perception/action effects for isolated events.

As a general rule, in this chapter I will simplify matters by considering action with respect to end point goals, in keeping with a dominant theme in motor cognition (Klapp & Jagacinski, 2011). For vocal production, motor planning is conceptualized as the synergistic coordination of all muscles in vocal motor control toward a pitch goal. The differing roles of various laryngeal muscles, or differences between phonation and respiration, are not considered here. On the piano, action planning will be considered with respect to spatial goals on the lateral plane (left to right), without consideration for what finger, hand, or arm movement is related to that plan. Of course, mapping of gestures and finger selection to piano keys involves many-to-one associations that are worth their own consideration. Be that as it may, many studies on the role of pitch content in piano performance to date have simplified this matter by creating melodies with only five pitches so that finger–key associations in the text are invariant. In practice, of course, there are many ways to produce the same pitch sequence on a piano, and the degree to which this figures into learning and perception/action associations is an important issue in its own right (cf. Meyer & Palmer, 2003; Palmer & Meyer, 2000).

Another simplification is that the present chapter focuses on studies of musically trained individuals and does not delve too deeply into differences across musically trained and untrained participants. Many of the studies summarized here include both groups, but because effects of training are worth deeper exploration I devote a whole chapter on training effects later in the book (Chapter 6).

COMPATIBILITY EFFECTS BASED ON PITCH–SPACE ASSOCIATIONS

Compatibility effects have been studied for a long time in psychology. The term *stimulus–response compatibility* was coined some time ago (Fitts & Seeger, 1953). Compatibility effects occur when perception/action associations are so strong that participants are unable to ignore the associations when they are irrelevant or counterproductive within a given task. A famous example is *Simon Interference* (Simon & Rudell, 1967; for an updated review see Hommel, 2011), in which the association between the hand used to make a response (right vs left) and the right/left location of an object on a computer screen influences how quickly a participant can categorize what kind of object is displayed (e.g., circle or square). Importantly, the categorization task has nothing to do with the spatial location of the object. Yet participants are reliably slower if the categorization response requires using the right hand (e.g., press right if you see a circle) and the object is displayed to the left, than if the object is displayed to the right. Presumably this happens because of the ideomotor mechanism discussed in Chapter 2. Because we are so used to associating rightward movements with perceptual events that appear to our right, it is difficult for us to plan rightward movements when we have just seen something to our left.

It is here that we find some of the simplest and most compelling evidence that pianists have learned to expect certain relationships between planned actions and pitch height. Consider a task in which you are asked to press the "P" key on the keyboard (on the right side) when you hear a violin timbre, and press the "Q" key (on the left side) when you hear a trumpet. The sounds you hear vary between a higher and a lower pitch, which makes no difference as to what response you select. However, pianists are significantly slower at making correct response if the spatial location of that response (e.g., a key on the right side of the keyboard) conflicts with how that spatial location is typically associated with pitch. Fig. 4.1 illustrates the kind of associations that have been explored. In English and many other languages, pitches are categorized using the verbal metaphor of "height," a metaphor that is further embodied in the structure of Western music notation. As such, perceiving a "high" pitch may be associated unconsciously with relatively high spatial locations on the vertical plane. In an experiment, therefore, participants may find it challenging to respond by tapping a spatially low key after hearing an acoustically "high" pitch. In doing this, the participant would have to counteract years of accumulated associations between pitch and space. On the piano, the most popular instrument learned in Western societies, a further association is developed in which acoustically "high" pitches are associated with the right side of the horizontal plane. As a result, it may be difficult for trained pianists to respond by pressing a key that is spatially left when hearing an acoustically "high" pitch.

Such tendencies have been documented in what was called the SPARC effect by Lidji, Kolinsky, Lochy, and Morais (2007), which stands for "spatial–pitch association of response codes" (another research group reported the same effect in the same year, but dubbed it

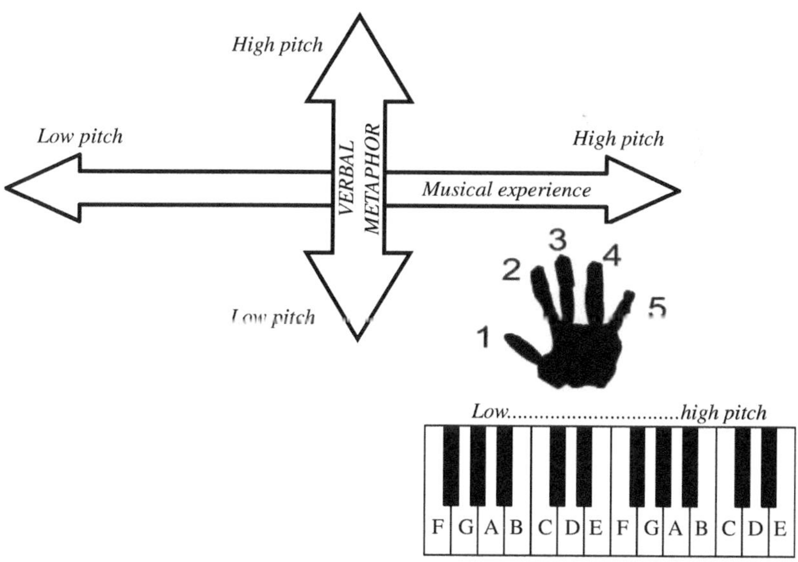

FIGURE 4.1 Associations between pitch and spatial locations based on experience.

the SMARC effect, referring to "music" as opposed to "pitch"; Rusconi, Kwan, Giordano, Umilta, & Butterworth, 2006). An analogous effect, called the SNARC effect, involves people's tendency to associate numbers with space (high numbers to the right), and was discovered earlier (Dehaene, Bossini, & Giraux, 1993). Nonmusicians show a SPARC effect, but only for associations in the vertical plane, which may be based on verbal associations (Lidji et al., 2007). Thus, musical training elicits a translation of this basic tendency to associate "high" and "low" pitch with "high" and "low" spatial coordinates. Similar effects are found when left/right movements are accompanied by task-irrelevant musical scales, with faster responding when the direction of a response matches the movement of the scale (e.g., a rightward response when an ascending scale is heard; Taylor & Witt, 2015).

Are such effects an artifact of the piano? Although the piano arguably has the most transparent mapping of pitch to space, most musical instruments may have some kind of way to map pitch height to dimensional changes in movement, as illustrated by the examples shown in Table 4.1. In general, all instruments may be influenced by sensorimotor constraints leading to instrument-specific coupling of perception and action, or "affordances" (De Souza, 2017; cf. Gibson, 1979). It is possible, based on these observations, that different kinds of musical training

TABLE 4.1 Mapping of Pitch to Movement and Space in Various Instruments

Instrument	Map to pitch height
Piano/Marimba	Lateral key position
Voice	Vocal fold tension
	Vocal fold thickness
	Subglottal pressure
Violin/Cello/Guitar	Lateral neck position
	String used
Trumpet/French Horn	Embouchure tense/relaxed
Trombone	Embouchure tense/relaxed
	Slide position
Clarinet/Sax/Oboe	Number of holes covered
	Embouchure tense/relaxed
Flute/Piccolo	Number of holes covered
	Breath intensity

cause individuals to map pitch to space in different ways. In keeping with this prediction, violinists show similar associations specific to mapping of space to fingers of their left hand (Kajihara, Verdonschot, Sparks, & Stewart, 2013).

Whereas the SPARC effect was based on mapping for single pitches, an analogous effect was identified earlier for a more complex kind of mapping: the mapping of several simultaneous pitches in a musical chord. Chords are particularly interesting because they can be performed differently on different instruments. A group of researchers from the Max Planck Institute for Cognitive Psychology (where the notion of common coding for perception and action was developed, see Chapter 1) set up an experiment in which pianists had to perform a musical chord as quickly as possible in response to a visual prompt, which could be verbal (e.g., "F major"), or abstract (e.g., a blue square, symbolizing F major). Along with this prompt, an irrelevant musical chord would sound that could match the chord represented visually or could match a different chord. Pianists were slower to respond if they heard an irrelevant mismatching chord, although nonpianists (who were trained to produce chords for the purpose of this experiment) did not show this interference (Drost, Rieger, Brass, Gunter, & Prinz, 2005). A subsequent study revealed that this effect was instrument-specific: When pianists heard irrelevant pitches in a guitar timbre, rather than a piano timbre, the effect disappeared. Conversely, guitarists only showed this effect when responding on a guitar to chords presented in a guitar timbre (Drost, Rieger, & Prinz, 2007).

An important question concerns how movement-specific these compatibility effects are. As noted earlier, my treatment of this area groups together spatial mapping with movement, as does my catalog of instrument examples in Table 4.1. Doing so, however, begs the question of whether such a simplification is appropriate! And so in the case of effects like SPARC, one can ask whether compatibility is based on mapping of pitch to arms, hands, and fingers used for movement, or whether this is more of an abstract mapping based on spatial dimensions. A research group in London, United Kingdom, led by Lauren Stewart, addressed this question in an experiment where the range of high and low pitches could be small enough to be producible on the piano with one hand without arm movements (participants responded with different fingers of their right hand) or were too large to do so (Stewart, Verdonschot, Nasralla, & Lanipekun, 2013). Compatibility effects were found in both cases, suggesting that the effect is based on more abstract mapping to space rather than something more movement specific. Other support for this view comes from the fact that the original effect was found to hold for musicians in general (not just pianists) and that it holds for changes in pitch height that are too wide to be performed on a keyboard within the spatial coordinates being used. At the same time Suzuki-trained pianists show sensitivity to pitch mapping based on how pitch height is represented on a violin (close to the body = high, far = low; Stewart et al., 2013). Ultimately, performers probably use some combination of action-specific and abstract mapping principles. An interesting case in point comes from expert cellists. Whereas an abstract coding of pitch to vertical space, based on metaphors used in language, would lead to associations of "high" pitches with higher positions in space, the mapping of pitch height to space on the cello is reversed. Cellists show no compatibility effects for pitch–space associations on the vertical plane, which suggests that these two sources of compatibility cancel each other out for this group (Lachmair et al., in press). Moreover, listeners instinctually associate pitch height with nonspatial variables, and various nonspatial metaphors for pitch are used across cultures (Eitan & Timmers, 2010). Thus, it is likely that various factors enter into the compatibility effects seen in these studies.

EFFECTS OF PITCH ALTERATIONS IN THE PERFORMANCE OF MELODIES

What happens when we consider action/pitch associations for whole melodies? Based on the previous section, one might predict that any alteration to pitch could disrupt performance of a melody, particularly for musicians and in particular for trained pianists. However, there are strong qualifications on what kinds of alterations to feedback content may disrupt production. This led to the conclusion for many years that feedback content may be completely irrelevant to production. At various points I have discussed the theory of Peter Howell (see Chapters 1 and 3) that feedback is only used to support the timing of execution. A major factor in this theory was the fact that alterations to the content of speech—for example, converting feedback from syllables to square-wave tones (Howell & Archer, 1984; see Chapter 7 for more discussion of this result)—lead to no change in the disruptive effect of Delayed Auditory Feedback (DAF). A comparable effect was reported for music performance by Finney (1997): well-learned performances of Bach inventions by pianists were unaffected by a variety of alterations to feedback pitch, in contrast to disruptive effects from DAF (see further discussion of learning effects in Chapter 2).

Two important factors likely contribute to these null effects. First, participants in altered feedback tasks already have a well-formed intention for what the performance should sound like, and thus can ignore altered auditory feedback. Why not, given that feedback in these cases offers no help to the performer? Second, melodies are patterns, and thus have Gestalt-like properties. Effects of AAF during melody production may therefore be based on properties of the entire sequence that is formed by auditory feedback, rather than mapping of pitch to space for individual events. These observations allow us to reinterpret the results just discussed. Manipulations of speech feedback by Howell and Archer effectively neutralized meaningful variability of content in the speech syllable. Finney's manipulations of pitch led primarily to "near-random sounding" melodies (Finney, 1997, p. 157).

A general theme in subsequent research is that the principles that lead to disruption of melody production may not be based entirely on specific associations for a musical instrument, but instead may mostly reflect our tendency to anticipate perceptual consequences from our actions that are reliably associated with the spatiotemporal pattern that forms the goal of action planning. In other words, when we plan to produce a melody, we do not plan a series of independent pitch events. Instead, as discussed in Chapter 1, we conceptualize these events as part of a hierarchically organized *pattern*. As a result, effects of altering feedback do not reflect properties of individuals as much as they do perception/action associations based on these overarching patterns. A useful analogy here is to Gestalt principles of organization that were developed for vision and have also been invoked as a way of understanding auditory perceptual organization (Bregman, 1990), and motor planning (Klapp & Jagacinski, 2011). In the case of manipulations like those used by Finney (1997) and Howell and Archer (1984), the lack of salient Gestalt-like properties may have made auditory feedback easier for participants to ignore.

The first hint that perception/action associations based on feedback content may be due to such organizational principles emerged in the early 2000s. Inspired by the fact that errors in piano performance often come from nearby events in the sequence (e.g., Palmer & van de Sande, 1993, 1995), a new manipulation of content within the software package FTAP (Finney, 2001a,b) was introduced in which keypresses would generate a pitch that was associated with another pitch from the sequence. In the most common version of this manipulation, which came to be known as a *serial shift* of auditory feedback content, each keypress would generate the pitch that the participant produced (or was expected to produce) one, two, or three events in the past or future, while being synchronized with the keypress (unlike the alterations discussed in Chapter 3).

The most common version of this manipulation is the lag-1 serial shift, in which each keypress generates the pitch associated with the immediately preceding keypress (just like one of the "misbehaving piano" examples given at the beginning of this chapter). An example is shown in Fig. 4.2. Thus, in this example, the participant would press the

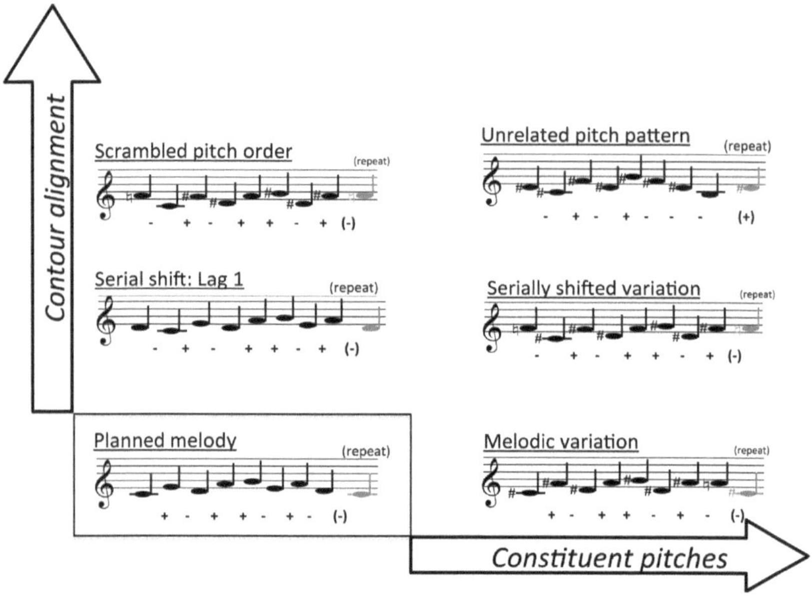

FIGURE 4.2 Examples of selected manipulations to feedback content, shown in music notation. One repetition of an example melody is shown, with *gray* notes at the end indicating the use of repeated cyclical performance. The melodic contour of each melody is shown below the notation, with " + " indicating upward pitch motion and "−" indicating downward pitch motion. Alterations from the planned melody can be roughly distinguished based on alterations to the melodic contour pattern (vertical dimension), and matches of the constituent pitch events (horizontal).

key for E4 (note 2), but hear C4 (note 1). When repeating the melody cyclically, which is standard in these experiments, one would hear the pitch of the last note in the melody (E4) when pressing the key associated with the first note (C4). Auditory feedback would thus be related to the planned sequence in a way that is similar to a musical canon, only at a lag that would not lead to the pleasant kind of musical relationships that one finds in a canon, where lags are typically much longer. In contrast to random-like alterations of feedback content, serial shifts do disrupt performance significantly, approximately doubling the proportion of pitches that pianists play incorrectly (Pfordresher, 2012).

A reasonable first reaction to this kind of an effect would be to question whether the effects of serial shifts from the past, which have been used most often, really differ from those of DAF. Reactivation of previous actions during sequence production typically indicates some problem of planning. Perseveration errors (those that involve repeating a past event) accompany performance that is generally sloppy and filled with errors (Dell et al., 1997; Drake & Palmer, 2000), and are a hallmark of neurological sequencing disorders such as aphasia (Dell et al., 1997), relative to anticipatory errors (playing something too soon). Models of sequence production typically involve a component that suppresses the activation of events once they are produced, with Node Structure Theory being a quintessential example (see discussion in Chapter 3). Feedback from past events may disrupt past events because they counteract this necessary suppression.

One can point to a very simple fact to undo this argument: Serial shifts from *future* events also disrupt performance. When feedback from piano keypresses is altered to generate pitches from *n* events in the future, disruption to performance is of a similar overall magnitude as for events from the past (Pfordresher & Palmer, 2006). To some degree, this consistency across past and future may reflect the structure of music. Although perseverations in music, like speech, indicate overall poor planning, it is also the case that musical performers usually make similar numbers of anticipatory and perseveratory errors, which is not like speech (Palmer & Pfordresher, 2003). Because pitches repeat often (we have only 12 pitches), suppression mechanisms used in speech models to turn off the past may not apply well to music (Vousden & Brown, 1998).

The fact that serial shifts disrupt production and randomly selected feedback pitches do not has several interesting implications for the Gestalt-like organizational principles that determine how performers associate action planning with the content of auditory feedback.

ROLE OF CONSTITUENT PITCHES

So far I have discussed two basic manipulations of content: serial shifts and the selection of feedback pitches at random (or near randomness). These reflect two extremes with respect to how similar the feedback melody is to the planned melody, as illustrated in Fig. 4.2, in which the notated example in the top right could be formed from randomly selected pitches. A serial shift, like the lag-1 shift shown in Fig. 4.2, presents a feedback melody that preserves all the features of the planned melody. All that changes is the sequential alignment of the feedback melody relative to the serial positions for the planned melody. By contrast, a feedback melody formed by randomly selected pitches will generally have little or no association with the pattern of the feedback melody, and as such the question of serial alignment is moot (representation of pattern by melodic contour, represented in Fig. 4.2, will be described in detail later). As such, one rather obvious difference between these manipulations is that serial shifts preserve all the pitch content of the original melody, whereas random or even "random-like" alterations do not. What happens when alterations to feedback pitch take on an intermediate level of similarity?

This issue was addressed in part by Gates and Bradshaw (1974), who found disruption similar in magnitude to DAF when participants performed a two-hand piano piece while hearing "extraneous" feedback described as "melody with accompaniment on the same instrument and in the same range as the test piece" (p. 107). It is possible that the extraneous melody used in this study was more similar to the test piece than the pitch sequences that resulted from the feedback manipulations used by Finney (1997), but both the absence of details in the older paper and the presence of many cross-experiment differences prevent us from drawing any firm conclusions.

A more systematic investigation of similarity in constituent pitches was reported in Pfordresher (2005). This paper contrasted effects of serially shifted feedback (based on a lag of 1) against randomly selected pitches and two other critical conditions. In one of these conditions, any succession of keypresses led to pitches from the planned (test) melody, but in a randomly permuted order. An example of this condition is shown in the top left of Fig. 4.2. As in serially shifted feedback, this condition caused participants to hear pitches they planned to produce, but with an unreliable serial relationship between the current serial position and the position associated with the feedback pitch. Perhaps more important, the resulting sequence of feedback pitches did not produce a melody with higher-order properties similar to the melody (like the unrelated pitch pattern in Fig. 4.2). The second new condition took each pitch from the melody, permuted its order, and then shifted each pitch up or down a semitone so that no single pitch matched any of the melody pitches. As a result, feedback pitches were not fully random but would cycle through a single set. However, that set would not match any pitches planned by the participant, and the resulting sequence also lacked the Gestalt properties of the planned sequence. In addition, sequences generated in this condition were not tonal.

Results suggest a graded effect in which feedback containing pitches from the planned melody, but in a scrambled order, led to errors that fell in-between those of serially shifted feedback (highest) and randomly selected feedback (no different from hearing normal feedback). Also, hearing a repeated cycle of pitches that do not match the planned pitches was indistinguishable from randomized feedback.

In discussing these results, I have referred in passing to "higher-order Gestalt-like" properties of melodies. What are they? There are several—all of which are standard hallmarks of music cognition research. By *higher-order properties* I mean organizational structures at a broad timescale that determine the ordering of individual events. These higher-order properties can lend to a melody the property of being a

Gestalt, that is, being perceived as a collective whole. As per the claims of classic Gestalt psychology, when an object has this property, its component parts lose salience relative to the whole. In the context of melodies, a Gestalt property would lead individual events to a secondary status of importance relative to higher-order properties that organize them into this Gestalt.

A critical hypothesis concerning the role of higher-order properties is whether the role of auditory feedback during the production of melodies may not be based on mapping of individual actions to single pitches but may instead be based on the coordination of the higher-order properties of the planned melody with higher-order properties of the melody generated by auditory feedback. If the content of auditory feedback is processed as a higher-order Gestalt, then it should be possible for feedback events to activate learned melodies as well as disrupt the online recall of melodies during performance. In Chapter 2 we discussed neuroscientific evidence for this hypothesis, in which listening to melodies one has learned to perform elicits motor activity in the brain (D'Ausilio et al., 2006; Lahav et al., 2007; Mutschler et al., 2007). Behavioral studies have addressed whether hearing a pitch sequence can facilitate access to that sequence.

Peter Keller and Iring Koch addressed this prediction using a response–effect compatibility paradigm (2008). Similar to the compatibility effects described earlier in this chapter, their focus here was on how compatibility influences the efficiency of access to a response. The main difference from the studies summarized before is that response–effect compatibility looks at how the *consequences* of produced actions influence later action selection, which is more like what happens in music. The authors used a music-like task in which participants had to recall a sequence of three button presses on vertically oriented keys (like a keyboard that's been stored on its side). Responses were based on neutral color plates (e.g., blue means tap "top," "bottom," then "middle"), and pitches that result from these keypresses were technically irrelevant to the task. In different phases of the experiment the mapping of pitch to space could be compatible (high pitch for the high key, etc.), or incompatible. But the mapping was always consistent within each of these phases. Keller and Koch found that, for pianists, responses were generally faster during phases with compatible mapping than incompatible mapping. The authors concluded that compatible mapping allowed the pianists to anticipate the consequences of their actions more effectively during the planning stage (after seeing the color-plate response cue), due to the ideomotor mechanism discussed in Chapter 2. Interestingly, the same effect was not found for nonpianists, even though the SPARC effect described earlier is found for nonpianists when pitch is associated with vertical locations in space and responses are explicitly associated with pitch height (as was the case in the Keller and Koch study). This difference may reflect different effects of stimulus–response versus response–effect compatibility.

Based on my own interest in these results, I joined Keller and Koch in a subsequent study that addressed whether altered feedback during a performance may activate action planning for a different learned melody (Pfordresher, Keller, Koch, Palmer, & Yildirim, 2011). We had participants (both pianists and nonpianists) learn and memorize two short melodies designed to be learnable by nonpianists (see Chapter 6 for further discussion). On each trial they would be cued to perform one melody or the other, and as they performed they would hear feedback from the melody they were playing or the "alternate" melody they had learned. Note that this is critically different from most altered feedback paradigms, which usually involve hearing irrelevant feedback or feedback that's related to the melody one is currently playing. If auditory feedback can activate an entire plan for a melody, then hearing alternate-melody feedback should make the retrieval of that melody easier. This is effectively what was found, although the support for the hypotheses arose in a rather roundabout way. During each trial, participants would hear an additional sound, which functioned either as a cue that they should immediately switch from the melody they were playing to the alternate melody, or it could be an irrelevant cue. The irrelevant cue condition was initially designed as a control so we could address what participants did simply as a result of hearing an additional sound as they performed. However, an insightful undergraduate assistant named Ece Yildirim (who became an author on the paper) noticed that participants were responding to this irrelevant cue. Specifically, when participants were hearing alternate-melody feedback, they would pause noticeably after hearing the irrelevant cue, as if it were difficult for them to suppress the inclination to switch melodies. And this ended up being the main result of the paper.

Of course, if auditory feedback during the performance of melodies is processed with respect to higher-order structures, then it should be possible to manipulate its effect by altering these structural properties. The simplest sort of higher-order representation in music cognition is *melodic contour.*

ROLE OF MELODIC CONTOUR

Melodic contour plays a surprisingly critical role in music cognition given its simplicity. Melodic contour is defined as the direction of a pitch change, independent of its magnitude (following Dowling, 1978). In Fig. 4.2, melodic contours for different altered feedback conditions are shown beneath music notation. When two successive pitches mark an upward change, the contour is coded as "+," indicating upward motion, whereas "−" means downward motion and "0" is unison (not true for any intervals in Fig. 4.2). However, there is no distinction between a very large upward change (e.g., first two pitches of "Somewhere Over the Rainbow") or a small upward change (e.g., first two pitches of "Do-Re-Mi"). In this respect melodic contour differs from how other contours, including visual contours, are defined.

Melodic contour represents directional patterns in pitch that may be highly salient to listeners. In performance, contour likewise represents salient changes in planned movement trajectories (e.g., shifts in direction on a keyboard, shift from tension to relaxation in the cricothyroid muscle for control of vocal pitch). Moreover, many studies support the salience of contour in music perception. Our first memories of a melody seem to be dominated by its contour (Dowling, 1978; Dowling & Bartlett, 1981; Dowling & Fujitani, 1971), and perceived melodic accents in music often happen at points when the direction of melodic contour changes (Boltz & Jones, 1986; Jones & Pfordresher, 1997; Pfordresher, 2003b; Thomassen, 1982). Composers use contour changes to convey accents (Huron & Royal, 1996).

It thus stands to reason that melodic contour may play a role in how we relate perception to action. Fig. 4.3 illustrates this association. If, for instance, movement to the right on a keyboard leads reliably to upward pitch changes, and movement to the left leads reliably to downward changes (as in the resulting feedback sequence shown in the upper part of Fig. 4.3), there is a correlation between changes to pitch height and planned movement transitions. Participants may respond positively to this kind of correlation even if they have no prelearned associations between specific positions on a keyboard and resulting pitches. By contrast, if there is an unreliable relationship between changes in movement and changes in pitch height (lower section of Fig. 4.3), interference may result as a consequence of this uncorrelated mapping. It is possible that sensitivity to these correlations is based entirely on directional changes in movement position and pitch. This kind of effect may be stronger than if auditory feedback in such a circumstance forms an upward interval (as anticipated) that is either too small or too large (e.g., pressing the "C" and "E" keys—a major 3rd—and hearing a full octave interval in auditory feedback). Moreover, the fact that serial shifts disrupt production may be because the melodic contour pattern in auditory feedback is a delayed copy of the melodic contour of the planned melody. So every planned directional shift comes out at the wrong time in auditory feedback.

If so, it should be possible to disrupt production by presenting auditory feedback that shares the same the melodic contour as serially

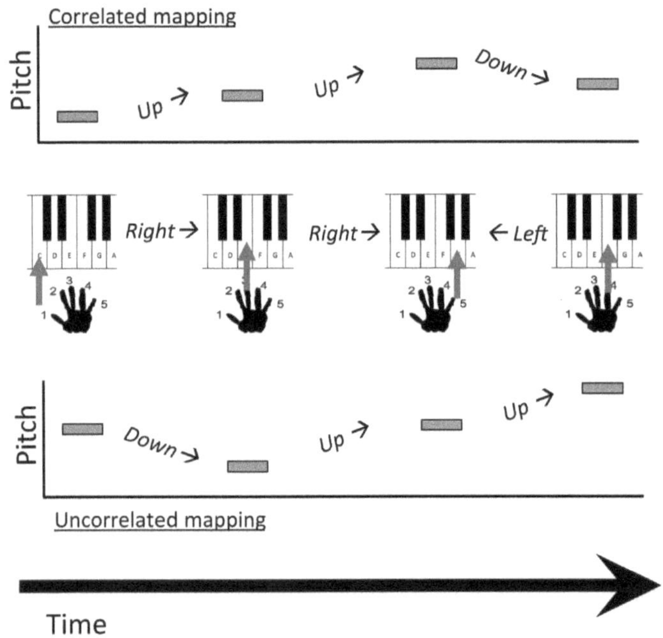

FIGURE 4.3 Schematic depiction of a sequence of keypresses on a piano over time (middle), and how these may lead to a feedback pitch sequence with changes in melodic contour that are correlated with movement (upper) or are not correlated with movement (lower). Auditory feedback is shown using "piano roll" notation in which rectangles stand for pitches that vary in time (*x*-axis) and pitch height (*y*-axis).

shifted feedback, but that differs in the pitches and melodic intervals of the original melody. Two studies (Jebb & Pfordresher, 2016; Pfordresher, 2008) tested this hypothesis by having keypresses generate a feedback melody that was some kind of melodic variation on the melody that participants planned. In one experiment the variation was simply a transposition of the original melody, whereas in a second experiment the intervals and pitches in auditory feedback were altered—leaving only the contour intact. An example of such a melodic variation is shown in Fig. 4.2. Pianists did not find this alteration from planned melody to melodic variation in itself disruptive, *unless* the melodic variation was serially shifted (also shown in Fig. 4.2). In other words, pianists responded to a serial shift of the planned melodic contour as opposed to a shift of the specific pitches from the planned melody.

A question that arises from this line of reasoning is whether disruption of the same magnitude may emerge when the feedback melody has a contour that has a perfect negative correlation with the planned melody. So every time the performer plans an upward pitch change, auditory feedback presents a downward change, and so on. Note that this kind of manipulation is not identical to a serial shift, which would only have a perfect negative correlation for a melody in which the contour direction switches for every other interval.

You may see now that we have come full circle: The initial SPARC effect manipulations are based on action/effect alterations that have a positive or negative correlation. A logical extension of this kind of manipulation to melody production would amount to an inversion of pitch mapping. For the keyboard, this would amount to a remapping in which movement toward the right lowers pitch and movement to the left raises pitch. An early study showed that pianists made significantly more errors when performing melodies with such a reversed mapping (Laeng & Park, 1999). A subsequent unpublished study from my lab, which included both pianists and nonpianists, suggested that effects of reversing pitch mapping are of a similar magnitude to serial shifts for both groups of performers (Pfordresher & Clafin, unpublished data). A complication in both studies is that music was learned from notation (although the notation was modified in the Pfordresher study, see Chapter 6 for further details), which for pianists is strongly associated with the typical mapping of pitch height to space (Stewart, Walsh, & Frith, 2004). A more recent study from my lab had performers learn melodies completely by ear, with no visual representation, and in this case disruption from a reversed keyboard among pianists was small in size (Pfordresher & Chow, in press). Ultimately more research needs to be done, but the melodic contour of auditory feedback appears to be an important factor in perception/action associations. However, melodic contour constitutes an incredibly coarse-grained description of musical pitch structure. Next I turn to a more complex, hierarchical representation of musical pitch: tonality.

ROLE OF TONALITY

Arguably, no feature of music has been the focus of as much interest as tonality (for extensive reviews, see Krumhansl, 1990; Krumhansl & Cuddy, 2010). The perception of tonality involves hearing a sequence of pitches and framing those pitches in the context of a musical scale. In so doing, the listener starts to perceive certain pitches as being dominant, given their role within the musical scale, and starts to form expectations for what notes ought to come next. Tonal relationships also play a major role in musical emotions because nondominant pitches lead to tension and more dominant pitches lead to relaxation, based on listeners' expectations. Thus, tonality involves perceiving pitches according to a hierarchy of importance.

A highly salient distinction for all listeners familiar with Western tonal conventions is the difference between "tonal" and "atonal" melodies. In contrast to tonal melodies, which constitute the vast majority of all music people listen to, atonal music comprises pitches that do not clearly map onto a single musical key. For most listeners, atonal music may sound unusual, disjointed, or even unsettling. Composers in the early 20th century embraced atonality as a new way to enliven music and to cast off previous conventions (which is a general theme of the 20th century). Unfortunately, audiences never quite got on board with these new developments. Schoenberg's prediction that people would eventually be heard whistling atonal music at the corner store (Cook, 1998) never quite came to pass. However, in the context of experimental psychology, the tonal/atonal distinction stands as an easy way to test how tonality may or may not constrain various aspects of perception and performance.

Consider melodic contour. As noted before, our first memory trace of a new melody is dominated by contour. But if we are encoding an atonal melody, contour is the *only* feature we retain. For tonal melodies it's a bit different: after their first exposure to a melody, listeners seem to recall both the contour as well as the scale along which the contour traversed, although not the precise intervals that were used (Dowling, 1978).

A roughly similar phenomenon holds for the way that performers respond to auditory feedback. There is an important qualification to a study I summarized earlier, which showed that the effect of serially shifted feedback may be based in large part on shifting of the melodic contour as opposed to specific pitch classes (Pfordresher, 2008). Specifically, serially shifted melodic variations did disrupt performance but only when the melodic variation was tonal. When pianists heard auditory feedback that formed an atonal melody with a melodic contour that matched the melody they planned to produce, they effectively "tuned out" auditory feedback. As a result, serial shifts of the atonal variation had no effect on performance.

There are two possible reasons why this might occur. In Pfordresher (2008), all planned melodies were tonal. Thus, the "tuning out" of atonal variations may have occurred because of a mismatch in what we might call the *tonal class* of planned melodies. Under this hypothesis, atonal feedback melodies do not interact with the planning of tonal melodies but may interact with the planning of an *atonal* melody. An alternative interpretation suggests that *tonal dominance* led to the absence of any disruption from atonal feedback melodies. Under this view, atonal melodies fail to form high-level Gestalts and because of that atonal feedback melodies, in any case, would not interfere with planning because such melodies would always be treated as extraneous.

The critical condition is one in which participants plan and produce atonal melodies. Jebb and Pfordresher (2016) had participants plan and perform tonal or atonal melodies in different experiments, while hearing feedback melodies that could be tonal or atonal melodic variations that in turn may be serially shifted relative to the planned melody. Results supported the tonal dominance hypothesis: regardless of the tonal class of the planned melody, only a serially shifted tonal feedback melody would disrupt production. In keeping with a general theme of this chapter, tonal melodies feature higher-order Gestalt-like properties that cause listeners to associated feedback melodies with planned melodies. Atonal melodies, by contrast, may lack these organizational properties and thus may dissociate from action planning more readily.

ROLE OF METRICAL ORGANIZATION

Here I will argue that melodic contour and tonality matter because they both contribute to *metrical organization*, which ultimately plays the most salient role in coordinating action planning with feedback content during music performance. Metrical organization refers to the kind of meter that is associated with a musical pattern, as well as the salience of that metrical pattern.

Readers with musical training will associate meter with bar lines and time signatures in music. These forms of organization are an important component in meter and have to do with the presence of a recurring time span (marked by the distance between two bar lines) that acts as a kind of organizational frame. And this organizational frame is not just important for the performer. As listeners, we implicitly pick up on whether a meter is binary or ternary. A binary meter, which contains a time span at some multiple of 2, will lead the listener to expect to hear patterns that recur at intervals that are multiples of two (think of the phrase from *Somewhere Over the Rainbow*: "Someday I wish upon a star and wake up where the clouds are far...," which features an alternating up/down pitch pattern). There is a second important component to meter, which is not so obvious from time signatures and bar lines. It has to do with the fact that meter determines a pattern of alternation between "strong" and "weak" beats and this pattern influences how listeners perceive the prominence of musical events (Palmer & Krumhansl, 1990). In other words, meter conveys a kind of accent pattern. This pattern is obvious when one learns to dance, where the pattern of foot and body movement one engages in has to be synchronized with the music's meter. This pattern of accenting is typically symbolized by a *metrical grid*. Examples of metrical grids are shown in Fig. 4.4 for a planned melody and auditory feedback that is serially

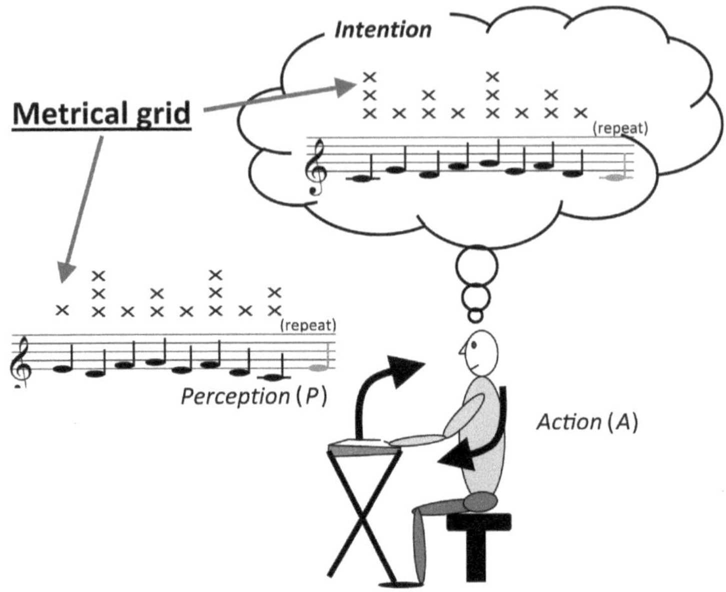

FIGURE 4.4 Illustration of metrical conflict caused by a lag-1 serial shift of auditory feedback. Notation for the planned melody (in the thought bubble) and auditory feedback are shown below metrical grids that represent alternating patterns of accent strength. Accented events fall beneath columns with more than one "x," with the strength of the accent proportional to the height of each column. Serially shifted feedback causes clashes between accents in the planned melody and accents in auditory feedback.

shifted by a lag of 1. Accents are related to the height of the column of "x's" above a given event. In meters like those shown in Fig. 4.4, accents alternate between weak beats (one "x") and accents of varying strength.

Where does meter come from? As a form of temporal organization, it is not surprising that meter often arises from rhythmic onset patterns (e.g., Essens & Povel, 1985; Povel & Essens, 1985), even rhythmic onset patterns that one may imagine (Palmer & Krumhansl, 1990). However, meter does not only reflect onset timing. Meter also results from melodic leaps, shifts in implied harmonies, and parallelism in phrase structure (Acevedo, Temperley, & Pfordresher, 2014; Jones, 1987). In particular, changes in contour create melodic accents, and the positioning of these accents helps to convey the metrical organization of music (Boltz & Jones, 1986; Ellis & Jones, 2009; Jones, 1987; Jones & Pfordresher, 1997; Pfordresher, 2003b). For instance, if melodic accents occur about every 4 beats, listeners are likely to perceive a binary metrical structure, whereas accents occurring about every three beats convey a ternary structure. Composers tend to position pitches that are salient in a tonal sense at metrical accents, suggesting a possible use of tonality to convey meter (Prince & Schmuckler, 2014). All of these pitch-based features can cause a listener to hear a kind of temporal structure, even in a musical pattern that is completely isochronous (i.e., containing no variability in note durations or interonset intervals). For instance, the meters associated with melodies shown in Fig. 4.4 cannot arise simply from patterns of onset timing, which are isochronous, but could be based on pitch patterns. Listeners use these temporal and pitch-based patterns to derive a perception of metrical structure, represented by a grid, which then functions as a structural frame or *mental schema* that—like tonality—helps to determine the salience of perceived musical events.

Meter is also important to the performer because it plays a role in planning of sequences (e.g., Palmer & Pfordresher, 2003). For instance, when performers make errors of serial order (see Chapter 1 for more discussion of serial ordering errors), those errors are more likely to originate from a sequence position that shares a similar metrical accent. Thus, given the metrical structure of the intended melody in Fig. 4.4, an error at position 3 (the note D), a strong metrical accent, would more likely arise from position 1 or 5 (C or G) because they are also associated with strong metrical accents. Errors like this imply that the performer has access to the produced event in the performance plan. The fact that performers' errors can be predicted by metrical accenting suggests that these accents are used as part of the planning process.

The fact that pitch can be used to convey meter offers a way to conceptualize effects of altering feedback content in a new way. If pitch structures contribute to temporal organization, it is possible to draw an analogy between the role of feedback content and the role of feedback synchrony. In both cases, we may think of the role of feedback as being temporally coordinated across perception and action, only at different timescales (global organization for content vs event synchronization) and perhaps by way of different neural pathways.

Consider what happens when auditory feedback is serially shifted. In such cases, the pattern of metrical accents implied by the pitch pattern in auditory feedback is displaced relative to where these accents were expected to occur. At the same time, the basic unit of temporal organization (e.g., binary or ternary) will be shared across the planned melody and the shifted auditory feedback. Fig. 4.4 illustrates how this works for a lag-1 serial shift. Whereas the first planned sequence event is intended to convey a strong accent, the pitch heard in auditory feedback is associated with a weak accent, and as the melody progresses, auditory feedback consistently counteracts the expected accent strength. We can refer to this as a phase shift of the metrical cycle associated with auditory feedback relative to the planned metrical pattern, just as we conceptualized effects of synchrony as resulting from the phasing of feedback onsets. To what degree does the disruptive effect of a serial shift result from phase shifting of metrical accents?

One way to address this possibility is to examine effects of different shift conditions, for the production of melodies having different metrical organizations. Because metrical accent patterns are periodic, some shifts will be more conflicting than others. A shift of one event is disruptive regardless of the meter associated with the planned melody. A shift like this leads to consistent mismatches between the accent associated with the planned event (for production) and the accent associated with the perceived event (in auditory feedback). However, a shift of two events will have different effects for different meters. If the planned melody has a binary meter with a 2-beat recurring cycle (e.g., a 2/4 time signature), this will result in a coincidence of relatively strong beats across perception and action, whereas a similar coincidence will not occur if the planned melody has a ternary structure.

In fact, this manipulation was the original idea behind serial shifts. The effect of this manipulation—an interaction between shift lag and planned meter—was pretty subtle (Pfordresher, 2003a). Across a sample of 21 pianists there was a slight tendency for serial shifts of lag 2 to be less disruptive when pianists planned a binary meter (for which a lag of 2 leads to matching metrical accents) and more disruptive when the planned meter was ternary. But the effect was only statistically significant for a subset of participants who had moderate levels of musical training. More to the point, serial shifts always increased error rates, regardless of their lag. So meter in this sense may be a contributing factor but it is not the primary factor in perception/action associations formed for melodies whose pitch structure reinforced the metrical accents.

One possibility is that alterations to the phasing of metrical accents may lead to effects that differ in only subtle ways. By analogy, participants have little trouble perceiving the meter of a pattern when melodic and rhythmic accents (e.g., lengthened notes) have the same period (e.g., every 4 beats) but are phase shifted relative to each other, whereas meters are more ambiguous when melodic and rhythmic accents suggest different metrical periods (Ellis & Jones, 2009; Pfordresher, 2003b). More than a decade later, a graduate student and I revisited the role of meter in auditory feedback (Pfordresher & Kobrina, 2017). This time, however, the auditory feedback melody could be based on either a meter that matched what the participants performed (e.g., a binary meter) or could match a different meter (ternary). The disruptive effect of hearing a melody with a different meter was of a comparable magnitude to hearing serially shifted feedback.

It is important not to forget that meter is more than just a feature of perceived music. Arguably meter has an even more important role for performers. At a basic level, bar lines help musicians parse a musical piece during music reading, and as such can influence performance timing and even recognition of melodies from notation (Sloboda, 1983). At a deeper, more abstract, level, meter acts as a reference frame for musical retrieval when notation is absent (long-term memory) and as a way to assemble plans, based on working memory, even when notation is present (Drake & Palmer, 2000; Mathias, Pfordresher, & Palmer, 2015; Meyer & Palmer, 2003; Palmer & Pfordresher, 2003; Pfordresher et al., 2007).

Therefore, if disruption from alteration of feedback contents happens because of incongruity in planned and feedback meters, the ability of performers to use meter as a planning frame may diminish as a result of such manipulations. Pfordresher and Palmer (2006) addressed this issue by analyzing patterns of serial ordering errors during performances with normal and serially shifted auditory feedback. In typical performances, serial ordering errors are influenced by meter: performers typically misplace the order of events by inserting them into other positions that have a similar metrical accent strength. Thus, if an error is made on a weak beat, and the performer substitutes a different note from the melody, that substitution is likely to come from another weak beat (Palmer & Pfordresher, 2003). Pfordresher and Palmer (2006) found that serial shifts diminish the influence of meter on the errors that pianists make. Pianists may therefore be less able to use meter as a guiding frame when the meter associated with auditory feedback interferes with the planned meter. In other words, serially shifted feedback may lead to confusion about the unfolding metrical organization of the sequence.

HIERARCHICAL PERCEPTION/ACTION COORDINATION: SERIAL SHIFTS AS METRICAL CONFUSION

Based on the evidence to date, I propose here a model of the role of auditory feedback content based on a form of coordination, in keeping with the broader Hierarchical Perception/Action Coordination (H-PAC) hypothesis introduced in Chapter 1. In Chapter 3 I proposed a similar view to address the role of feedback timing. However, coordination of content differs in some important respects and thus may reflect a different hierarchical level. I propose that sequential coordination of perception and action occurs based on meter, and use as a starting point an extension of the range model of planning (Palmer & Pfordresher, 2003; Pfordresher et al., 2007). In this model, the accessibility of events surrounding one's current position in a melody is determined by the product of two factors that lead to activation of sequence events in working memory.

The first factor is serial proximity (S):

$$S_x = a^{|x|/t} \tag{4.1}$$

where x indexes the distance of the event from the current position. Referring to my earlier example from Fig. 4.4, if the current position is 3, events from positions 1 and 5 would both have $|x| = 2$. The t parameter comes from performance tempo (beat period duration), and a is a free parameter that estimates a baseline level of short-term memory capacity (how much information is lost per unit of time). This equation predicts an exponential drop-off in event activation with distance that is steeper when short-term memory capacity (a) is lower and also steeper when one performs at a faster rate (lower t), as illustrated in Fig. 4.5A.

The second component, more significant for our present purposes, comes from the aforementioned effect of metrical organization on planning:

$$M_x = 1 - \frac{1}{n}\sum_i \frac{|m_i - m_{i+x}|}{m_i + m_{i+x}} \tag{4.2}$$

Lowercase m values refer to the metrical accent associated with the current position i and x refers to events at a serial distance of x events from that position. Values of m come from the metrical grid. For

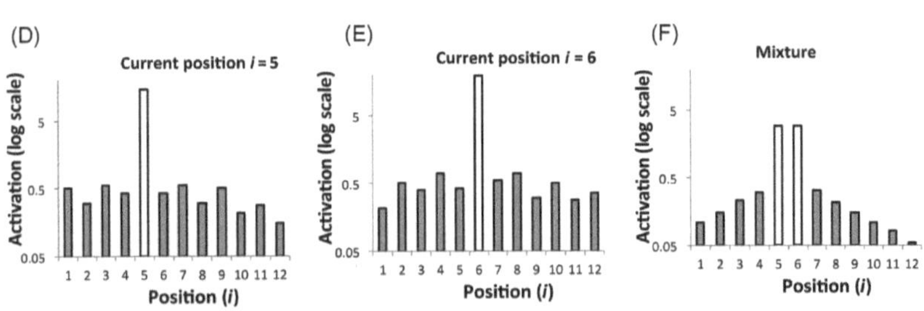

FIGURE 4.5 The range model of planning. (A–C): Interaction of serial (A) and metrical (B) components, leading to predicted activations based on their product (C). (D–F): Positional uncertainty resulting from a serial shift in which the activation gradient associated with auditory feedback (D) lags one position behind the planned current event (E), leading to a combined gradient that poorly disambiguates the correct event from surrounding events (F).

instance, if you are at event 1 of the melody depicted in Fig. 4.4, $m_1 = 3$. The accent for a distance of one away is $m_2 = 1$, and for a distance of two is $m_3 = 2$. The heart of this equation is the difference between accent strengths associated with surrounding events. When this difference is small, as in m_1 vs m_3, events share a similar metrical accent strength that leads to relatively high activation of the event at distance x from the present position. The resulting pattern of activations alternates between strong and weak based on metrical associations of the present event with surrounding events, as illustrated in Fig. 4.5B. The range model predicts that event accessibility in planning comes from the product of serial and metrical activations, illustrated in Fig. 4.5C.

Whereas Eqs. (4.1) and (4.2) account for activations of events surrounding the current position, the range model predicts that these activations may facilitate correct retrieval of the current event. As noted elsewhere, individual pitch events in music have little meaning outside their surrounding context. Thus, our ability to retrieve musical events in a sequence likely benefits from the presence of surrounding activations. An extension of the range model incorporated this logic by predicting that accurate retrieval is proportional to the square of activations from surrounding events (Pfordresher et al., 2007).

How does this framework account for effects of altering feedback content during a musical performance? A general theme is that serial shifts happen when the metrical content of the feedback melody is (a) perceptually salient and (b) not aligned with the planned melody. One possibility is that such conflicting perception/action relationships lead to *positional uncertainty* in performance planning. One role of meter may be to act as a cue for one's position in the sequence; when the feedback meter conflicts with the metrical code associated with the present position, uncertainty may result. During a lag-1 serial shift, the pattern of activations in planning (Fig. 4.5D) conflict with the pattern associated with planning (Fig. 4.5E).[2] As a result, contextual information necessary for production fails, leading to retrieval errors and less salient meter. This result can be modeled by assuming that serial and metrical activations during performance with shifted auditory feedback come from the product of activations based on the planned event and the event associated with auditory feedback. The product of these joint activations may reflect interactions across perception and action. A prediction based on this assumption, shown in Fig. 4.5F, exhibits reduced activations for surrounding events, with a weaker metrical influence, lower activation for the current event, and ambiguity concerning the present position.

[2]Note that the use of a log scale in panels D–F, which is necessary given the high activation of the current event, diminishes the visual salience of serial distance.

Although this framework is preliminary, it can account for increased error rates brought about by serially shifted feedback as well as the fact that serial ordering errors during serial shifts exhibit a weaker influence of metrical associations than errors made during normal auditory feedback (Pfordresher & Palmer, 2006).

NEURAL REPRESENTATION

How is feedback content processed in the brain? As a starting point, we may expect that the brain should recruit ideomotor or "mirror neuron" areas discussed in Chapter 2, including the inferior frontal gyrus ("Broca's area," and its homolog in the right hemisphere). However, studies to date suggest that more areas come into play when the brain has to coordinate action planning with alterations of feedback content. This issue is of considerable theoretical importance. Consider Peter Howell's theory of feedback processing, detailed in Chapters 1 and 3. According to this view, auditory feedback is processed as an extraneous signal. If so, neural responses to altered feedback should differ from responses to errors that a performer actually makes. However, if feedback is processed in a way that is integrated with action planning, as others have proposed, neural responses to altered feedback may be similar to responses based on errors that a performer actually makes.

One way to address this point is to measure the latency with which the brain responds to these alterations. Research using *event related potentials* (ERPs), as discussed in Chapter 1, can be used to measure neural responses to a specific event of interest with high temporal precision. For now, that event of interest would be some alteration in auditory feedback. A general theme of ERP research is that brain responses take longer if they involve higher-level abstract processes, and are earlier for more basic sensory and motor processing, which have reflex-like properties. A particularly critical response that the brain shows is its response to making an error, referred to as error-related negativity (ERN). This is a shift in polarity in the negative direction that occurs about 100 ms after a participant makes an error (Gehring, Goss, Coles, Meyer, & Donchin, 1993). When the processing of an error is based only on response feedback (in the absence of producing an error), the resulting component is referred to as a feedback ERN (also called FRN), with a longer latency of about 250 ms (Ridderinkhof, Ullsperger, Crone, & Nieuwenhuis, 2004). Clemens Maidhof, as a master's student, decided to find out how similar neural responses to alterations of auditory feedback were to responses to errors that pianists actually make (Maidhof, Vavatzanidis, Prinz, Rieger, & Koelsch, 2010). Comparisons between ERP responses to errors pianists made, with normal auditory feedback, and alterations of feedback pitch on correctly produced notes yielded some interesting similarities and differences. Whereas the ERP signal produced a robust negativity around 200 ms after the feedback alteration, similar to other studies showing a feedback ERN, no such response was found when pianists actually made errors. Rather, self-made errors led to neural responses *before* the error was made (cf. Herrojo-Ruiz, Jabusch, & Altenmüller, 2009; Maidhof, Rieger, Prinz, & Koelsch, 2009). So your brain may respond to altered feedback as a signal that something is wrong, but perhaps not in exactly the same way as an error you made.

A more recent ERP study focused on an earlier response, the N1 (Mathias, Gehring, & Palmer, 2017), which reflects early sensory processing of auditory onsets. Typically the N1 is reduced for expected auditory feedback such as that consistent with one's own plan, presumably because one's forward model suppresses the perception of stimuli that are consistent with the model (Poonian, McFadyen, Ogden, & Cunnington, 2015). Mathias and colleagues explored whether this suppression of N1 is reduced when the content of auditory feedback is altered, which would suggest that feedback content is not interpreted as being self-produced. The authors tested effects of isolated serial shifts (i.e., shifts on a single event surrounded by other sequence events with normal feedback), from the immediately preceding as well as the very next (future) pitch event in the sequence. Consistent with the view that this type of altered feedback is interpreted as extraneous (not selfproduced), the N1 increased when feedback was altered, but only when the alteration was from a future event. Mathias et al. (2017) proposed that this result may reflect participant's greater ability to suppress activation of planned as opposed to future events (cf. Dell et al., 1997).

To date, the only study to identify the location of neural responses to altered contents, using fMRI, is one by Pfordresher et al. (2014). Trained pianists performed melodies from memory while listening to normal or altered auditory feedback, including a condition in which the pitches of all feedback events were serially shifted by a lag of 1. The supplementary motor area of the brain responded more to this condition than during performance with normal auditory feedback. This area was also observed in studies of acquired sensorimotor associations (Fig. 2.2) and responses to asynchronous feedback (Fig. 3.7). Thus this area seems to have a general role in sensorimotor associations of many sorts. Another area that responded to serially shifted feedback was area SPT, which also figures in studies of sensorimotor associations for the vocal system (Hickok et al., 2003; see Fig. 2.2).

Two other areas that were active during serially shifted feedback have important implications for the role of feedback content. One particularly interesting area is the anterior cingulate cortex. This area, which is beneath the surface of the midcentral frontal cortex is thought to play a role in error monitoring, and may register the presence of two competing response possibilities (Carter et al., 1998). The fact that this area responds to serially shifted feedback reintroduces the prospect that the brain treats altered content as a kind of an "error," or at least as feedback that is inconsistent with planning. The second area was the inferior parietal lobule, which may play a role in processing melodic transformations (Foster, Halpern, & Zatorre, 2013; Zatorre, Halpern, & Bouffard, 2010). Taken together, the activity in these two areas suggests that pianists process serial shifts as disruptive transformations of planned melodies.

Somewhat surprisingly, the inferior frontal gyrus, which is found in many studies of feedback processing, did not respond to serially shifted feedback. We had anticipated activation in this area given studies discussed in Chapter 2, suggesting that this area serves as a hub of the mirror neuron system for music production (e.g., Lahav et al., 2007). The absence of activity in this area runs counter to the idea that auditory feedback content is processed as self-generated.

CONCLUDING REMARKS

In this chapter I summarized research on how people associate the planning and execution of musical sequences with the resulting sequence of pitch contents. I also reflected on the theoretical implications of these results that affect the role of higher-order Gestalt-like organization in music, the organizational role of meter, and the higher-order influence of tonality. In the next chapter I reflect on differences in the role of feedback synchrony (discussed in Chapter 3) and content with respect to a possible shared representation for perception and action in music.

CHAPTER

5

Timing Versus Sequencing in Music

The last two chapters have discussed the role of temporal synchronization and sequential organization in auditory feedback. There are several reasons why I separated the discussion into two chapters. The most critical reason is that results to date suggest that associations between perception and action based on synchronization rely on different mechanisms than those that are based on serial order. Temporal synchronization of sound with action relies on basic timekeeping mechanisms. In Chapter 3, I proposed a nonlinear oscillator mechanism, which seems to account for the available data. In Chapter 4, I proposed that conflicting metrical organizations, mediated by the presence of salient Gestalt-like organizational principles (e.g., tonality) lead to positional uncertainty that hinders serial recall. At the same time, there is a common theme across both chapters: in neither case is it necessary that auditory feedback events actually represent the melody one attempts to produce. Asynchronies can disrupt performance even if content is altered or even neutralized (e.g., Howell & Archer, 1984), and feedback with a conflicting meter can disrupt sequencing even if the feedback melody shares no pitches in common with the planned melody. Thus, although the results reported in this book all have to do with perception/action associations, they do not require explanation by a closed-loop system that is based on feedback principles in the literal sense (discussed in Chapter 1).

The role of this chapter is to reflect further on the implications of these different results, relate this distinction to the broader psychological literature, and to discuss in more detail the degree to which a full dissociation between sequencing and timing mechanisms is warranted by the data. This leads to an elaboration of the Hierarchical Perception/Action Coordination (H-PAC) model (first introduced in Chapter 1, and further developed in Chapters 2 and 3), in which music performance

Sound and Action in Music Performance
DOI: https://doi.org/10.1016/B978-0-12-809196-8.00005-X

relies on a shared representation with distinct hierarchical levels for sequencing and timing. Sensorimotor integration during music performance abides by this separation.

Although it is common to separate sequencing and timing, some clarification and critical evaluation of this separation is necessary. Here is the way in which I conceptualize the distinction, based on how this terminology is used elsewhere, as well as observations from data summarized in this book.

Timing: Refers to the production and perception of event onsets in real time, irrespective of the content of what is produced and the sequencing of events differing by category.
Sequencing: Refers to the production and perception of distinct event categories, and the overarching structure into which these sequences fall, distinguished by their sequential order.

These definitions are consistent with applications in the existing literature, which generally associate timing with rate and sequencing with the likelihood of one event following another, as well as sequential dependencies (Krampe, Mayr, & Kliegl, 2005; MacKay, 1987; Matheson & Sakata, 2015; Zatorre et al., 2007).

WHY THE SEPARATION?

At the outset, it is worth considering why it is useful to distinguish sequencing from timing in the first place. After all, what we call sequencing and timing both effectively involve time. For a series of events—words in a sentence, or notes in a melody—to unfold in the appropriate order the events have to be separate in time. It may be possible to conceptualize the production of serial order as a kind of "ordinal" measurement scale for time, that is, one that follows the rank ordering of events one after the other but not the degree of temporal separation.

As it turns out, there are several reasons to consider sequencing and timing as distinct from each other, whether or not they reflect truly independent neurocognitive processes. First, both features can provide critical information to the perceiver, as is the case in music. If you change the timing of a familiar melody, for instance, it may become unrecognizable, even if the serial order of pitch events is correct (Hébert & Peretz, 1997). Second, sequencing and timing may be weighted differently as one learns an instrument. Whereas beginning pianists are more likely to make timing than sequencing errors, the reverse is true of more accomplished pianists (Drake & Palmer, 2000). Third, although some tasks require accuracy of both sequencing and timing, others do not.

Consider a typical experiment on memory for serial order. All that matters in such a task is that the list is recalled in the correct order—timing means nothing. Conversely, rhythm production tasks in music cognition often make negligible demands on memory for serial order.

In keeping with these observations, psychologists have long proposed some kind of separation between sequencing and timing. Karl Lashley can be credited with introducing this distinction to mainstream psychology, in an influential paper that primarily addressed the shortcomings of behaviorist ideas of sequence production (Lashley, 1951). Patterns of sequential behavior, such as the errors people make when speaking, suggest that retrieval is based on the encoding of hierarchical structure and not, as behaviorists had proposed, simple associative chaining of one event to the next. Moreover, an individual can produce a rhythmically regular sequence and yet exhibit disruptions of serial order; people do not always stammer when they produce a slip of the tongue. Various subsequent theories followed up on these initial insights. Although timing, like sequencing, is often considered to be hierarchical, rhythmic hierarchies may constitute a distinct representation such that altering rhythms may influence the meaning of a particular sequence (Krampe et al., 2005; MacKay, 1982, 1987; Martin, 1972). As mentioned earlier, this sort of timing hierarchy may draw on sequencing resources by virtue of the informational content involved.

The main contribution here is to extend the distinction between sequencing and timing to the way that perception and action are coordinated in complex tasks like music performance. As such, the present proposal expands on important theoretical antecedents, in particular those of Donald MacKay and Peter Howell. Although Donald MacKay has proposed a theory of perception and action based on a division between sequencing and timing (MacKay, 1987; see Chapter 1, for more discussion), his theory was created before results from altered auditory feedback experiments suggested that such a division would be useful for understanding online perception/action associations. As such, the ideas in this chapter can in large part be considered an expansion on the proposals of Node Structure Theory. Likewise, Peter Howell's views on the role of auditory feedback acknowledge this sort of division (see Chapter 1), although as with MacKay the role of auditory feedback is considered to be more limited than I propose here.

The division into sequencing and timing also ties into an important issue within music cognition: the differing roles of pitch and time. Within a musical sequence, successive events vary on multiple dimensions, with pitch and time being the most critical for our experience of music (recognizing the tune, understanding its emotional basis, etc.). Each successive event is distinguished with respect to pitch height, with the sequencing of pitches often being referred to as the *melody.*

In addition, the relative timing of successive onsets may be said to determine the *rhythm* of the sequence. This distinction between melody and rhythm maps onto what I refer to here as sequencing and timing. Although the dimensions of pitch and time are integrated in the auditory signal, evidence suggests that the brain processes melody and rhythm separately, at least at early processing stages. Cases of acquired and congenital amusia exhibit dissociations of rhythmic and melody processing (Hyde & Peretz, 2004; Marin & Perry, 1999). Among normally functioning populations, melody and rhythm may contribute independently to recognition (Hébert & Peretz, 1997), and perception of phrasing (Palmer & Krumhansl, 1987). However, in other tasks, such as meter perception (Ellis & Jones, 2009) and accent perception (Jones & Pfordresher, 1997; Pfordresher, 2003b), melody and rhythm influence each other in an interactive way. A similar question figures into the present consideration of how sequencing and timing operate in the coordination of perception and action.

THE ROLES OF SEQUENCING AND TIMING IN SENSORIMOTOR MUSIC ASSOCIATIONS: THE HIERARCHICAL PERCEPTION/ACTION COORDINATION MODEL

Fig. 5.1 illustrates the two basic levels proposed here as they relate to the production of individual events. As can be seen, "timing" in this model knows nothing about the planned content for any given serial position. The role of timing is simply to designate critical points in time. These points in time may designate when an action plan will be executed, for production, or may involve the temporal orienting of attention, for perception (cf. Large & Jones, 1999). Note that this role for timing is very much like the role of timing nodes in Node Structure Theory. The theories diverge in their accounts of how timing mechanisms mediate perception and action in performance, as discussed in Chapter 3.

The representation of sequencing in Fig. 5.1 represents the output of higher-order serial recall processes; in Chapter 4, I proposed the range model of planning as a framework on which to base predicted activations (Palmer & Pfordresher, 2003; Pfordresher et al., 2007). At any point in time, an event associated with the current position receives a higher level of activation than surrounding events, whose activations vary as a function of their separation from the present event, as well as perhaps metrical associations as discussed in Chapter 4. The figure illustrates how this would work for a simple 4-note sequence [A4 B4 C4 D4]. At the first position A4 receives highest activation, with decreasing

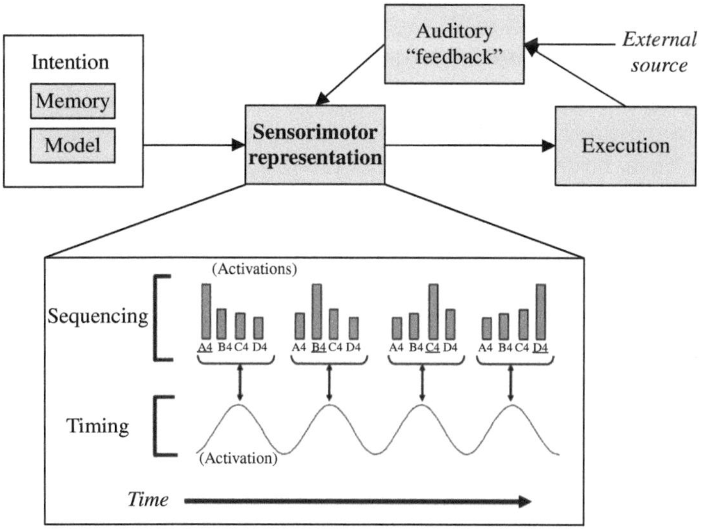

FIGURE 5.1 The sensorimotor loop relating perception to action, including the hierarchical structure of the sensorimotor representation. Source: *Based on Pfordresher and Kulpa (2011).*

activations to surrounding events. As we progress through the sequences we find peak activation shifting to later events. During performance, this evolving distribution of activations determines what event will be produced at each serial position. During perception, these activations can reflect anticipated sequence positions but more critically represent the categorization of perceived events into pitch classes.

Errors in this system can arise from various sources, including but not limited to alterations of auditory feedback. Timing, for instance, may be noisy, leading to imprecise production of rhythms or lapses of attention. At the level of sequencing, activations form a probability distribution for action selection. Thus at any point in time there is a chance that the selected event could be one that does not have a maximal activation (cf. Pfordresher et al., 2007). Likewise, in perception errors can occur based on miscategorizing pitch events or failing to track the timing of events.

The representation shown in Fig. 5.1 is presumed to serve both perception and production, based on various lines of evidence summarized in previous chapters. In this context, disruptive effects of altered feedback occur because mismatches between perception and action disrupt the dynamics of this representation. I here outline a series of claims arising from the manner in which sequencing and timing are thought to relate in this model.

Claim 1: Sequencing and Timing Have Different Roles in Perception and Action

The differing roles that these levels have leads to a basic but critical finding that I have not highlighted before: different alterations of auditory feedback have different effects on production. Whereas asynchronous feedback significantly disrupts performance timing, most prominently by slowing down performance, its effects on the accuracy of sequencing are not significant as long as appropriate controls are used (more on this later). By contrast, alterations to feedback content—when they are disruptive—increase errors while not affecting performance timing. I have referred to this pattern of results as a behavioral *sequencing/timing dissociation* (Pfordresher, 2003a; Pfordresher & Kulpa, 2011; Pfordresher & Mantell, 2014; Pfordresher et al., 2014; Pruitt & Pfordresher, 2015), suggestive of an architecture like that shown in Fig. 5.1.

Fig. 5.2 summarizes the results from several studies in which the same group of participants produced melodies while hearing either

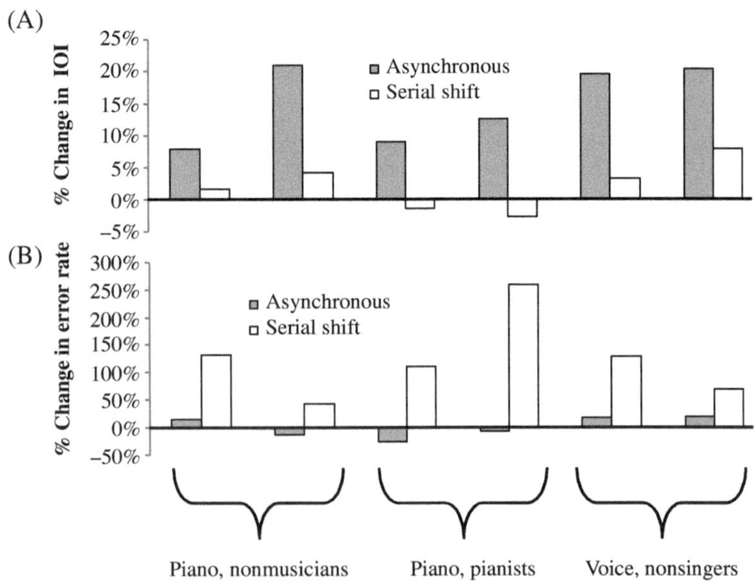

FIGURE 5.2 Summarized results from studies measuring effects of asynchronous (A) versus serially shifted (B) auditory feedback within-subjects. *Y*-axes represent differences between altered and normal feedback conditions, relative to performance in the normal condition. From left to right, data sets are: Couchman et al. (2012), Experiment 1; Pfordresher and Mantell (2012), keyboard trials; Pfordresher et al. (2015) and Pfordresher (2003a), Experiment 4; Pfordresher and Mantell (2014), singing trials; Pruitt and Pfordresher (2015), singing trials.

asynchronous auditory feedback or feedback for which the pitch content is serially shifted feedback by a lag of 1 (see Chapter 4, for more detail) on different trials. A highly similar pattern of results can be found across all studies, despite differences in musical training (pianists, non-musicians) and the motor system used (fingers and hands, voice). Specifically, each study demonstrates a dissociation. Asynchronous feedback increases the mean of interonset intervals (IOIs) (tempo slows while having little effect on accuracy). By contrast, serially shifted feedback increases error rates while having little effect on produced timing.

Claim 2: Perception and Action Interact; One Does Not Guide the Other

A second important tenet of this theory is that the relationship between perception and action in music reflects basic sensorimotor coordination tendencies and not the use of feedback for monitoring and control. In this respect the H-PAC model resembles the EXPLAN model, which also does not consider auditory feedback to be used for control purposes. The major difference between the models is the purported role of feedback content. The H-PAC model allows for the use of feedback content for perception/action coordination and thus in this sense performers do use feedback content. Critically, the role of content is contingent on the degree to which the higher-order pattern formed by successive auditory feedback events conveys a salient temporal structure. This structure may then be associated with patterns of movement based on task-specific or task-general associations, which will be discussed further in Chapter 6. Interference happens when the temporal structure of auditory feedback counteracts the planned pattern of movements.

It is important to note that the present theory does not preclude the use of auditory feedback for error monitoring purposes. Certainly that often happens. However, auditory feedback does not typically serve this purpose and does not have to serve this purpose. Thus, when error monitoring occurs the performer must bring in additional executive control processing. I will address the role of error monitoring in relation to the role of auditory feedback in Chapter 8.

A critical prediction arising from Claim 2 is that the role of auditory feedback for solo performances should be comparable to effects of auditory information from other agents. For example, information from other performers in an ensemble may be disruptive to a performer if its relationship to one's performance plan is difficult to coordinate. In fact, this claim was central to Peter Howell's observations about the effect of DAF, as discussed in Chapter 3. However, he focused only on onset timing. The H-PAC model takes the broader view that coordination in pitch patterning as well as onset timing may influence a performer. Thus, patterns involved in canons and fugues may prove challenging to a performer in a way that resembles the effect of serially shifted auditory feedback. Moreover, practicing one's part before a concert could be difficult if someone is performing a potentially conflicting pattern nearby. I discuss the role of auditory information in multipart performance in Chapter 9.

Claim 3: Sequencing and Timing Have Distinct Neural Representations

If there are distinct networks for coordinating perception and action based on sequencing and timing, then it stands to reason that they should be based on distinct neural networks. This possibility was addressed in the neuroimaging study by Pfordresher et al. (2014), which has been mentioned in earlier chapters. In that study, trained pianists performed novel melodies from memory while brain activity was measured using fMRI. During these performances, auditory feedback was varied in several ways, including a normal baseline condition, asynchronous feedback, and a lag-1 serial shift.

Several brain areas were more active when people performed with altered auditory feedback, in comparison to brain activity in performance with normal auditory feedback. Three of these areas were active during both asynchronous and serially shifted feedback and thus may be considered as reacting to the disruptive effect of alterations in general. One area was the supplementary motor area, extending into the premotor area, which is often associated with motor planning. The cerebellum was also active during both altered conditions, although there was more widespread activity in this area during serially shifted than asynchronous feedback. The cerebellum may serve various functions, including timekeeping and forward modeling of perceptual feedback based on action planning (discussed further in Chapter 4). Finally, the insula was active in both conditions. This area is beneath the surface, in a fissure (fold, separating the frontal, temporal, and parietal lobes). The insula does not make much of an appearance in many introductory treatments of cognitive neuroscience, yet it may have a critical role in sensorimotor integration. Recent research on singing suggests that this area responds to alterations of auditory (Kleber, Friberg, Zeitouni, & Zatorre, 2017) and somatosensory (Kleber, Zeitouni, Friberg, & Zatorre, 2013) feedback, the latter referring to a study in which participants' vocal folds (a.k.a. vocal chords) were anesthetized! The fact that this area also responds to alterations of auditory feedback in piano performance suggests it has a general role in relating perception and action beyond any specific motor system.

More important, other areas were differentially responsive to alterations of synchrony versus content. The right inferior frontal gyrus responded to asynchronous feedback but not to alterations of feedback content. This is the right homolog to Broca's area, and figures prominently in studies concerning the development of perception/action associations discussed previously (Bangert & Altenmüller, 2003; Lahav et al., 2007; Mutschler et al., 2007). These results seem to contradict each other, if one assumes that effects from learning studies are based solely on content. In fact, my coauthors and I were initially puzzled by our results from the asynchronous condition. However, it is important to note that the learning studies did not measure brain activity during altered feedback (even Bangert & Altenmüller, 2003, who manipulated feedback during training). Thus, it is possible that the frontal activations seen in those studies were influenced by the temporal associations between perception and action.

Several areas were differentially responsive to serially shifted feedback. A particularly critical area was the junction of the temporal and parietal lobes within the Sylvian fissure, the so-called area SPT. Many studies from the domain of speech suggest that this area plays a critical role in associating two neural networks: a ventral (lower) path used for comprehension, and a dorsal (upper) path for motor planning (Hickok & Poepple, 2007). Thus, area SPT may play a critical role in perception/action associations, and respond to their sequential dissociation during serially shifted auditory feedback.

Another theoretically interesting area that responded to altered feedback content was the anterior cingulate cortex (ACC). Studies designed to befuddle action selection typically lead to activity in this area. For instance, say you have an experimental task to respond when you see a right-pointing arrow (→), but not a left-pointing arrow (←), and only to respond to an arrow that's positioned in the middle of an array. What happens when a leftward arrow is surrounded by rightward arrows? Like this:

→ → → ← → → →

Your response, of course, should be to withhold responding, but all those rightward arrows will incline you to respond anyway. This is called the "flanker task" (Eriksen & Eriksen, 1974), and is often used to test people's ability to exert control over action selection. Thus, the ACC may respond to conflicting response tendencies, which may be part of why serial shifts disrupt the sequencing of actions (Carter et al., 1998).

A third interesting area that responded to serially shifted feedback was the inferior parietal lobule. The parietal lobe is commonly considered to be important for locating objects in space. Although it is possible that this activation reflects confusion about spatial mapping on the keyboard, more recent research suggests that the parietal lobe plays a role in recognizing reversals of melodies, and thus may play a more general role in constructing mental transformations of sequential information (Albouy, Weiss, Baillet, & Zatorre, 2017; Zatorre et al., 2010). Thus, activation in this area may respond to the fact that altered feedback content involves a kind of sequential transformation of the planned melody. It bears noting that this area also responded to randomizations of feedback pitch in a different condition.

Taken together, this evidence supports a neural basis for the dissociation of sequencing and timing in perception/action coordination. Unfortunately only one study to date has observed the effects of both these manipulations within the same sample. However, the presence of ACC and cerebellar activations during serially shifted feedback was recently replicated in a study using occasional alterations of feedback pitch (Herrojo Ruiz, Maess, Altenmüller, Curio, & Nikulin, 2017).

Claim 4: Sequencing and Timing Have Different Timecourses

Based on the underlying functions involved, it stands to reason that effects related to the synchronization of auditory feedback with action should be associated with different temporal dynamics than effects related to feedback content.

As discussed in Chapter 3, the synchronization of perception and action influences a central timekeeper that is designed to regulate the phase and period of rhythms that we perceive or produce. The effect of asynchronous feedback on production appears to influence phase, which responds instantaneously to perturbations. Likewise, the disruption of a timekeeper's phase alone—independent of its period—do not last. In keeping with this prediction, when feedback asynchrony is removed timing goes back to one's original rate. In addition to its rapid appearance and disappearance, the effect of asynchronous feedback is fairly homogeneous when it is present, as can be seen in Fig. 5.3A. This figure presents a selection of data from Pfordresher and Kulpa (2011), which was discussed in Chapter 3. Note that the magnitude of disruption does not increase or decrease over time while feedback is altered, nor are there noticeable periodicities in its effect. Rather, the effect of asynchronous feedback approximates a step function in the timing of IOIs across the sequence. The effect of asynchrony also does not require repeated presentations of the delay: even a single asynchrony within a melody will cause a temporary perturbation of timing (Furuya & Soechting, 2010).

By contrast, feedback content is thought to disrupt higher-level plans for retrieval based on global properties of the sequence one produces,

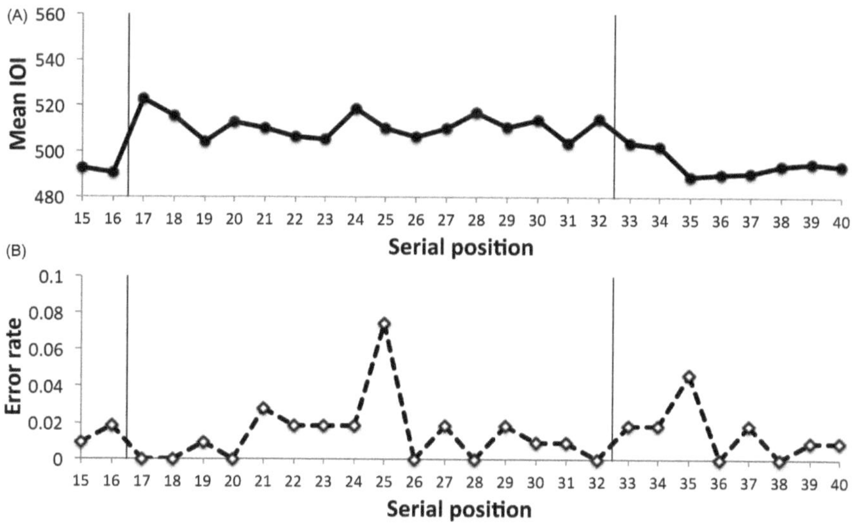

FIGURE 5.3 The timecourse of disruption from asynchronous feedback (A) and serially shifted feedback (B). In each panel, vertical guides delineate a sequence of tones for which auditory feedback was altered. Events before and after this region were associated with normal auditory feedback. Each data point is an average across participants. The first 16 events of the trial functioned as a baseline; only the last two events from this phase are shown as reference points against which to judge the effect of alterations. Source: *Data are from Pfordresher and Kulpa (2011).*

as described in Chapter 4. These mechanisms involve associations of content with serial position that evolve more slowly over time, and thus may take longer to progress and likewise may last longer. Another experiment from Pfordresher and Kulpa (2011) addressed the timecourse of disruption from serially shifted auditory feedback. Fig. 5.3B illustrates the differences in the effects of these feedback alterations. In contrast to the asynchrony effects, it takes several repeated experiences of serially shifted feedback before any measurable disruption occurred, after participants experience serially shifted feedback on four events. Furthermore, once disruption does occur, the effect is not homogeneous across sequence positions. We see a sharp upward jump when participants reach the position associated with the beginning of the 8-note sequence, at position 25 in the trial, and highly variable disruption thereafter. This pattern likely reflects the fact that sequential coordination between perception and action are based on higher-order organizational factors, such as the position at which the sequence repeats itself. As such the beginnings of sequences constitute critical points in the sequence. Finally, unlike asynchronous feedback, serially shifted feedback's effect does not immediately vanish.

Thus, asynchronous feedback has an effect that is immediate, homogeneous, and transient, whereas serial shifts have an effect that is cumulative, periodic, and lasting. The finding that serial shifts have a cumulative effect corresponds to results in several other studies that include feedback alterations on individual isolated events, such as a single event within an entire trial. Typically these kinds of alterations do not disrupt sequencing of action, but instead lead to temporary perturbations of timing. For instance, the ERP study of Maidhof et al. (2010), discussed in Chapter 4, found no changes in accuracy or timing following feedback alterations. It should be noted, however, that the musical patterns in that study were different from those used in many other studies. Pianists played scales and arpeggiated chords with both hands, and at each point both hands played a full octave. Thus, each successive pitch involved a change in hand position, whereas in most studies successive pitches involved selecting a new finger. Furuya and Soechting (2010) reported a small number of pitch errors when the pitch of a single note was altered during performances of Bach pieces, although the authors did not report any statistical analyses of these errors, suggesting that the rates were too low to be reliable. In contrast, Mathias et al. (2017) reported a statistically significant increase in errors following isolated feedback alterations. Unlike the other studies, Mathias and colleagues used serially shifted alterations along with alterations from unrelated pitches. Thus, manipulations in this study overall may have led to more interference with planning. Thus, it is possible to elicit errors with alterations to single pitch events, although this effect may not be as reliable as the effect of isolated temporal perturbations.

Pfordresher and Beasley (2014) looked further into the cumulative effect of alterations to feedback content, by varying the frequency with which altered pitches were heard by participants during a trial. Thus, some trials included alterations on only one pitch event, or on all pitch events, as had been done in the past, whereas other trials included alterations on 25%, 50%, or 75% of pitches. In contrast to Pfordresher and Kulpa (2011), altered pitches were not always presented on a series of contiguous events (except of course when every produced pitch was altered), but instead were presented at unpredictable intervals. The proportion of produced errors increased as a function of the proportion of altered events. This result verifies that the cumulative effect of altered pitch content reflects the frequency of all events that are altered, and is independent of how easily the performer can predict whether the next produced action will yield normal versus altered feedback. It is also important to note that Pfordresher and Beasley (2014) only found these results when altered pitches were serial shifts based on the produced melody. When altered pitches were selected at random, alterations did not increase error rates, regardless of the frequency with which the alterations were experienced.

However, there is an important qualification on this distinction between sequencing and timing. Whereas the effect of serial shifts on keyboard production is cumulative, as in the studies described above, the effect on vocal production may be more immediate. As noted previously there are many more studies of piano performance than of voice (and even fewer for nonpiano instruments); two studies that addressed the effect of serially shifted feedback on singing found rapid effects of altered pitch, with a much greater effect magnitude than found for keyboard production (Pfordresher & Mantell, 2012; Pruitt & Pfordresher, 2015). This rapidity likely reflects the online control of pitch in vocal production, a matter I return to in Chapter 7.

Claim 5: Sequencing and Timing Differ in Role of Sequential Complexity and Movement

Another claim arising from the purported difference between sequencing and timing is that the regulation of timing should not rely greatly, if at all, on the complexity of the sequential pattern one produces. If the performance plan involves an isochronous series of onsets, effects of asynchronies should be present regardless of whether the series is of a single repeated tap (a task often used in the timing literature), or a melody. To a first approximation, this claim holds. Effects of DAF, as noted in Chapter 3, have been found in a wide variety of tasks ranging in complexity from simple isochronous tapping (e.g., Aschersleben & Prinz, 1997), to the recitation of excerpts from Lewis Carroll's *Alice in Wonderland* (MacKay, 1968).

What does matter for effects of timing is the relationship between movement trajectories and the timing of asynchronous feedback onsets. This idea stems from observations made some time ago by Howell et al. (1983), who attributed the disruptive effect of DAF to the presence of "...a rhythmic signal with a similar pattern to the activity being organized but out of synchrony..." (p. 774). One way to interpret the "activity being organized" is as a movement trajectory. As discussed in Chapter 3, the effect that asynchronous auditory feedback has on performance is influenced by the state of one's movement trajectory at the time the delayed feedback sounds.

In a more general sense, however, it is important to note that the magnitude of disruption from asynchronous feedback, for the delays that happen to be most disruptive, does not depend on task. In other words, playing a more complex melody may not make you more susceptible to disruption by asynchronous feedback than when you engage in simple tapping. Although the pattern of disruption can vary across movement type, the magnitude of disruption does not. Most important, there is no evidence to date that a more complex task increases the disruptive effect of asynchronous feedback when the role of content is controlled (e.g., Pfordresher & Benitez, 2007; Pfordresher & Kulpa, 2011). It is important to note that studies of traditional DAF do sometimes show increased disruption with sequential complexity (e.g., MacKay & Burke, 1972), but as discussed in Chapter 3, it is possible that these effects are influenced by occasional disruptions to feedback content that can occur when using traditional DAF in which delays are based on absolute time.

It is a different story for effects of alterations to feedback contents. Here, complexity seems to matter. A rather trivial, but useful, observation makes this clear. Consider again an isochronous tapping task. Alterations to pitch do not disrupt performance, although quite interestingly some subtle effects to timing can happen based on the implied velocity based on pitch changes (Ammirante & Thompson, 2012). Furthermore, informal observations in my laboratory suggest that the performance of overlearned patterns like musical scales are immune from disruption by altered pitch feedback, at least for pianists who are proficient enough to treat these patterns as overlearned (cf. Maidhof et al., 2010). In short, there needs to be some kind of demand on serial retrieval in the first place for alterations of content to disrupt sequencing.

Beyond this, do manipulations of the degree of complexity influence the magnitude of disruption? One comparison that affirms this speculation can be found in Fig. 5.2B. Note that the effect of serial shifts varies considerably across the two groups of pianists. These two groups performed different melodies. The pianists in Pfordresher et al. (2015), on the left, performed pieces with invariant key-to-pitch mapping. This group performed in an fMRI scanner and thus could not observe their own hands; we therefore thought requiring changes in hand position would be unnecessarily burdensome under the circumstances. By contrast, pianists in Pfordresher (2003a), on the right, performed melodies that did require changes in hand position and thus were given a more challenging task.

Another unpublished series of studies in my laboratory addressed this issue within a single group of nonpianists (Pfordresher & O'Sullivan, unpublished data). In one half of the experiment, participants would memorize an 8-note melody as in many other studies, and then perform it with auditory feedback that was normal, serially shifted, or that featured a randomly selected pitch. In the other half of the experiment, participants would complete the same procedure with a more complex 12-note melody. The effect of serially shifting auditory feedback was greater for the 12-note melody than the 8-note melody.

The influence of musical training has differing influences on the role of sequencing and timing as well. This could be discussed here as *Claim 6*. However, Chapter 6 focuses entirely on effects related to musical training, so I will wait to discuss this claim from the model until that chapter.

THE RELATIONSHIP BETWEEN SEQUENCING AND TIMING

A major issue within cognitive science for more than 30 years has been the question of whether cognitive functions have a property known as *modularity* (Fodor, 1983). In a simplistic sense, modular functions operate independently from each other. Of the various features Fodor identified as characterizing modularity, two are most critical and both are of interest in this research area. One feature, *domain specificity*, refers to the kind of input a module accepts. Because modules operate independently, they are designed to operate on different subsets of information from the environment. Within music cognition and speech perception, a major question has been whether cognitive functions we use to process music and language function in a domain-specific way. If so, then processes that we use to comprehend spoken language may be separate from those that we use to recognize a melody based on properties other than song lyrics. I will address this aspect of domain specificity in more detail when I compare the role of sound for music and speech in Chapter 7.

The other critical feature of modularity, which for Fodor was its *sine qua non*, is *encapsulation*. Encapsulation refers to whether information processing within a module is influenced by processing within other modules, or by knowledge about how that process works. By this logic, if sequencing and timing are modular, then disruption of sequencing, by a manipulation like the serial shifting of auditory feedback or a pitch shift of feedback, should operate independently of any kind of disruption to timing.

Thus far my discussion of sequencing and timing may lead one to conclude that I consider them to be modules. However, to address this question ideally, one cannot simply fall back on studies that manipulate synchrony and content in isolation from each other. It is necessary to observe the effect of altering both at the same time. A modular approach would predict independent or additive effects in such cases. If, for instance, an asynchrony slowed down IOIs by 100 ms and a serial shift increased errors by 5%, then presenting both manipulations should lead to both slower and more errorful production. When I have described these kinds of situations to people in casual conversation, most people seem to think something like that would happen. They typically chuckle in a way that communicates their sympathy for participants who would be made to take part in such torture.

In fact, when both sequencing and timing manipulations are combined, their effects are under-additive. For instance, in my dissertation I found that the effect associated with serially shifting feedback was cut in half when the shifted feedback was also asynchronous (Pfordresher, 2003a). An even more extreme reduction in the effect of asynchronous feedback on timing emerged when the asynchrony was combined with a serially shifted feedback pitch. Finney (1997) reports a similar result. As discussed in Chapter 4, Finney did not find any disruptive effects when feedback pitch contents were altered in a random-like way, whereas DAF did disrupt performance. These findings appear to support Howell's claim that feedback content is irrelevant to the fluency of performance. However, contrary to Howell's claim Finney also found that the effects attributable to asynchronous DAF were no longer significant when he combined DAF with a random-like alteration of pitch. Thus, pitch was relevant to Finney's effects, though not in a way that most would have predicted.

Thus the existing evidence runs counter to the notion that sequencing and timing constitute separate modules for representing sequential structure. In fact, it is possible that sequencing and timing do not even constitute *representations*. Representations, as the term suggests, are abstractions that "stand for" some kind of environmental stimulus. The formation of a representation involves recoding the original stimulus into a new form that may be easier for the brain to retain, retrieve, and process. These notions have long been associated with the "computer metaphor" of the mind although there are varying degrees of commitment to this metaphor even within traditional cognitive science.

A radically different approach to the mind has emerged that extends from dynamical systems theory, or DST (e.g., Kelso, 1995; Port & Van Gelder, 1995), also discussed in Chapter 3. DST approaches to cognition emphasize the role of selforganized temporal patterns based on coordination of internal and external states, in contrast to the idea that we comprehend input by assimilating it to a static representation. Not surprisingly, dynamical systems approaches flourished first within motor control, and studies of motor coordination. In such contexts, the importance of real-time coordination is transparent. However, a few moments of reflection on the phenomenology of perception can lead one to reconsider whether perception, too, involves coordination in time. Our visual experience of the real world typically involves active scanning of a scene that evolves in time. Even more so, listening to music involves such obvious temporal coordination that people sometimes find it hard to suppress rhythmic body movements while listening (Janata, Tomic, & Haberman, 2012).

As such, sequencing and timing may not so much reflect different representations of structure as they do different *modes of coordination*, particularly given the apparent relationship of sequencing to higher-order temporal properties of pitch like contour and the communication of meter. By modes of coordination I refer here to the sense that coordination can be based on different timescales and different kinds of environmental contingencies. Consider a task like dancing to an old Motown song (Burger, London, Thompson, & Toiviainen, in press). Different body parts will synchronize to different periods within the music based on their physical properties. Trunk movements will likely entrain to moderate periods like the beat, whereas arm movements may reflect faster periodicities, based on their physical properties. All body parts are coordinated with music, while manifesting different timescales within the environmental pattern.

THE ROLE OF RHYTHMIC PATTERNS

Some readers will note to this point that my reference to "timing" in the H-PAC model fails to distinguish two senses in which timing exists. The studies discussed in this book so far for the most part involve the production of isochronous sequences, with the production of timing being related primarily to how well one can maintain the tempo. These measures focus specifically on the role of absolute time in music (see Chapter 3, for further discussion of absolute time). Although tempo is important, a great deal of research on musical time focuses on *rhythmic patterns* in music that may be based on relative time. Rhythmic patterns, to some degree, exist independently of tempo: you can sing a familiar tune ("Pop Goes the Weasel") at a wide range of tempos and it still sounds like the same rhythm. To what degree does "timing" as described in the model relate to rhythmic patterns?

I bring this point up here because research on the role of auditory feedback has not explored well enough how perception and action are coordinated with respect to rhythmic patterns. Thus, my comments here have to be considered speculative. Rhythms pose a challenge because in a rhythm we have a serially ordered sequence in which the content is defined by time itself. In this respect rhythms involve a blending of timing and sequencing. In fact, Krampe et al. (2005), in a study of sequencing and timing for rhythmic finger tapping, framed rhythm production as "sequencing" and only tempo as "timing." As such, the planning of rhythmic patterns may involve interaction between timing and sequencing. Following other leading theories of rhythmic organization (e.g., Jones, 1976; Jones & Boltz, 1989; Martin, 1972), I propose that rhythms are based on temporal hierarchies based on nested oscillators tuned to different frequencies. These periodicities help to establish an overarching temporal framework for rhythms. Input from sequencing may be necessary to determine whether a given onset is marked by an event or not. Thus, rhythms themselves may be based on interactions between sequencing and timing.

A particularly informative study regarding rhythmic patterns is Finney and Warren (2002). As discussed in Chapter 3, participants in this study tapped a rhythmic pattern, and DAF elicited insertion errors when delayed feedback onsets occurred at about the time when the rhythm would usually have a "rest" instead of an onset. Within the tentative framework described here, these errors may reflect the way in which feedback coinciding at a temporally salient point can elicit sequencing behavior.

CONCLUDING REMARKS

In this chapter, I described in more detail the distinction between sequencing and timing laid out in the H-PAC model, consistent with recent research on effects of altered auditory feedback on music performance. Sequencing and timing are distinguished by their effects on performance as well as underlying neural and cognitive representations. At the same time, sequencing and timing do not function independently, as in a modular system, but instead interact.

The first five chapters of this book have focused on the general role of sound in music performance, and in discussing these general effects I have argued for an architecture that serves to coordinate perception and action in sequential behaviors, including but not limited to music (H-PAC). The remaining chapters of this book focus on how the effects described so far are modulated by different experimental contexts and individual background, as well as the theoretical implications of these changes. In the next chapter, I focus in more detail on how the amount of musical training an individual has influences the role of sound in music performance.

CHAPTER

6

Effects of Musical Training

In this chapter I address an important issue that has been lingering in the background of all the previous chapters: to what degree does musical training influence the way performers associate action planning with associated sounds (auditory feedback)? I have withheld discussion until now for a couple reasons. First, the basic question driving this chapter is related to a much deeper issue concerning how we acquire associations between perception and action and perception, referred to here as *sensorimotor learning*. This topic warrants discussion in its own right. Second, discussing specific aspects of auditory feedback in more detail previously allows me to reflect on these various features in an integrated fashion here. Similar to Chapter 5, this chapter functions not so much as a review of the literature, most of which I have already covered in previous chapters, but more as a reflection on how existing results inform a theory of how sensorimotor learning in music may occur. Moreover, some of the ideas presented in this chapter have not yet been tested to my knowledge, and as such this chapter indulges somewhat in speculations for future research to address.

My own foray into the area of musical training was motivated by both theoretical and practical issues. On a theoretical level, by the time I was in graduate school (the latter half of the 1990s), many studies of how musical training affected perception existed (cf. Smith, 1997). However, the study of music performance—always of modest scope compared with perception—had focused almost exclusively on expert musicians. From a certain perspective this makes sense: if one assumes that a musically untrained person has no capacity for musical expression except after considerable training, studies of "nonmusicians" become highly impractical. However, placing this limitation prevented those of us who studied performance from addressing the same kind of interesting questions that were being probed in perception. For instance, does musical training simply involve fine-tuning the muscles, or do

Sound and Action in Music Performance
DOI: https://doi.org/10.1016/B978-0-12-809196-8.00006-1

trained performers actually conceptualize music differently (e.g., Palmer & Meyer, 2000)?

Practical concerns then forced me to address this long-standing theoretical question. In 2003 I found myself engulfed in the assistant psychology professor's nightmare: I found myself unable to recruit enough trained pianists to complete a series of experiments that were underway. The University of Texas at San Antonio, my employer at the time, has a terrific music school but my attempts to capitalize on their talent were not bearing fruit, and the tenure clock kept ticking. So I started to reconsider what is involved in playing the piano. Psychology is rife with studies in which participants press buttons in a sequence (e.g., Keele, Ivry, Mayr, Hazeltine, & Heuer, 2003). To some degree, playing the piano is a highly refined button-pressing skill. Furthermore, one of the biggest challenges in learning to play the piano is to master the many-to-one possible mappings between fingers and piano keys (i.e., a single piano key can be pressed using any finger). This challenge can be eliminated simply by creating melodies based on the first five notes of the C-major scale (all white keys), thus allowing participants to map each finger of their right hand to a single invariant piano key. Playing the melody can be conceptualized simply as a sequence of finger movements. Finally, rather than having nonpianists learn the abstract system of notation used in common practice, I designed simplified notation that explicitly relates to finger patterns as shown in Fig. 6.1, derived from notation being used for motor imagery experiments in Caroline Palmer's lab at the time.

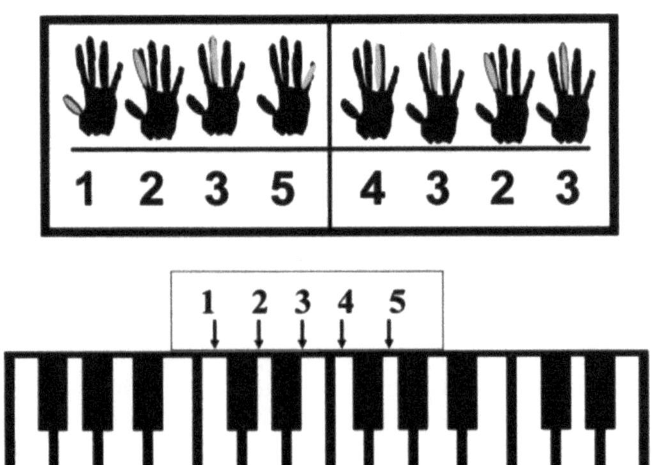

FIGURE 6.1 Notation system used for nonpianists. Source: *Reprinted from Pfordresher (2005).*

I want to highlight a subtle but important change of focus in this chapter from most of the preceding chapters. The primary concern of preceding chapters has been on relationships between perception and action during the performance of a known melody. Although part of Chapter 2 considered the role of feedback presence during learning, to this point we have not considered the broader picture surrounding the process of *sensorimotor learning:* the formation of associations between perception and action that allow us to acquire a specific skill. Sensorimotor learning in music, by this definition, involves learning the contingencies that pertain to a specific musical instrument, rather than learning a particular melody on that instrument. Certainly, when we learn to play a melody on the piano for the first time we also engage in sensorimotor learning. But the sensorimotor learning that goes on has to emerge out of the more specific learning that comes from the melody in question. In this chapter I consider how basic principles of sensorimotor learning relate to the role of auditory feedback across individuals who differ in their acquired musical skill.

PRINCIPLES OF SENSORIMOTOR LEARNING

What kind of potential does the musically untrained individual have for acquiring sensorimotor associations? Is this individual a *tabula rasa* (a blank slate), as would follow from a radically empiricist view of learning? If so, instrument configurations should matter little, and effects of manipulating feedback for untrained individuals should be negligible in comparison with effects found for individuals with musical training. Or do performers begin musical training with constraints, inclinations, and facilities that they carry over from their broader life experience? Note that the questions posed here do not simply regurgitate the classic "nature versus nurture" debate. The issue here has to do with how much sensorimotor learning transfers from one task to another.

Of course, most individuals within a culture acquire some explicit knowledge of how instruments are configured even if they have never performed on such an instrument. In my own lab, we have informally quizzed participants how piano keys relate to pitch height. Most participants with no musical training know that high pitches are mapped to the right, and the few that do not initially report knowing this fact suddenly remember it after pressing just a few keys on the keyboard in almost every case. As such, the really critical issue from the standpoint of musical training effects is not the role of explicit knowledge about perception/action associations. Instead, we are concerned with sensorimotor associations formed during learning. Sensorimotor associations involve neural representations that are distinct from exclusively perceptual or motor processing (cf. Chapter 2). Simply knowing in an explicit way about what actions lead to high or low pitches may not engage these representations.

In Chapter 2 we reviewed research in which musical training led to significant changes in behavior and brain activity even over a short timespan. For instance, neuroimaging studies that involve sensorimotor training show significantly greater activation in motor areas during perception tasks after this training (Chen et al., 2012; Lahav et al., 2007; Mutschler et al., 2007). One interpretation of such results is that there were no such motor associations prior to learning. Similarly, behavioral data from compatibility tasks, discussed in Chapter 3, show significantly larger compatibility effects for trained musicians and often no significant effects of this sort for untrained participants (e.g., Lidji et al., 2007; Rusconi et al., 2006). Such results could lead one to accept a radically empiricist view of sensorimotor learning—that no sensorimotor motor associations exist prior to learning.

It may be unwise to accept this radical empiricist view too readily. First, although musically untrained people may not have sensorimotor associations specific to a musical instrument, all individuals have a wealth of sensorimotor experience that may influence musical associations. In other words, nonmusicians have not acquired *task-specific* associations for music, but may have *task-general* associations acquired in other tasks when they engage in learning. This idea not new: Jean Piaget's theory of development, for one, follows from the claim that sensorimotor contingencies that one learns early in life provide the backdrop for the development of abstract sophisticated logical schemas that are based themselves on abstract contingencies. Thus, when learning a musical instrument, one's progress may be shaped by the abilities and biases that have been acquired through general life experience. Experience thus may lead to *constraints* on sensorimotor learning that in some cases may lead learning to be more efficient, while in other situations may lead to difficulty.

I propose three critical constraints for sensorimotor learning in music. The first is *synchrony*, which is associated with temporal aspects of perception/action associations. When we produce an action we expect the consequences to coincide with meeting the goal. Associative binding between perception and action thus can occur when the two are synchronous, but may be difficult or impossible when actions are not synchronized with perceptual feedback. The importance of synchrony for learning has been recognized since Aristotle described temporal contiguity as a primary factor in associations formed during learning. Fig. 6.2 illustrates how this kind of binding might occur. Based on the oscillatory timekeeping framework introduced in Chapter 3, we might predict that the tendency for actions to be associated with perceived

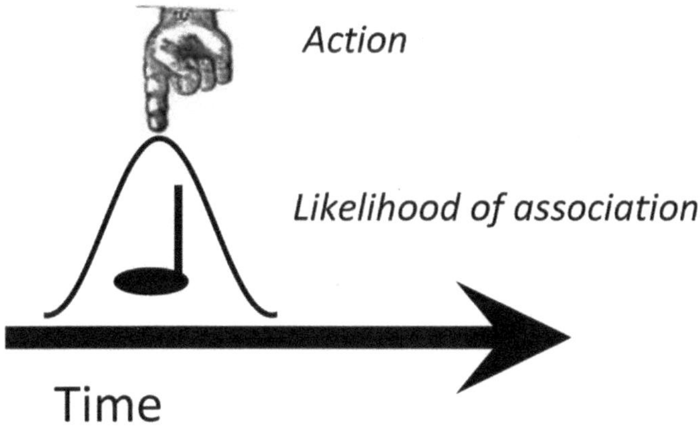

FIGURE 6.2 Temporal synchrony constraint on sensorimotor learning.

consequences might vary with respect to synchrony based on a sinusoidal function, with the strength of associating decreasing nonlinearly as a function of time. Note that this framework also predicts that associations based on synchrony may be based on the associations of rhythmic patterns between perception and action.

The role of synchrony in sensorimotor learning was addressed in a study by Brown and Palmer (2012). In that study, participants learned melodies while hearing auditory feedback that was either perfectly synchronized with their actions (as on a standard keyboard) or that was "weakly coupled" with actions. In the weakly coupled condition, participants performed along with a recording of an ideal performance. As a result, the slight nuances and perturbations of timing that the performer may have incorporated would not be matched in the recording, leading to occasional subtle discrepancies of timing. Later recognition of practiced melodies was worse for some participants after they performed with weakly coupled feedback than when they performed with perfectly synchronous feedback. However, this effect was not strong or reliable across participants, possibly relating to the subtlety of the manipulation.

The other constraints I consider here primarily bear on mapping of actions to pitch content in the previous literature. Fig. 6.3 presents three illustrative examples of how mapping from piano keys to pitch could vary, all of which have been used in existing studies. Normal mapping on the piano appears as a point of reference for the two mappings that involve alterations (inverted and scrambled mapping). I will refer back to these examples in the ensuing discussion.

A second task-general constraint on sensorimotor learning is *reliability*: individuals may expect the immediate perceptual consequences of actions to be consistent from one instance to another. Quite simply, if

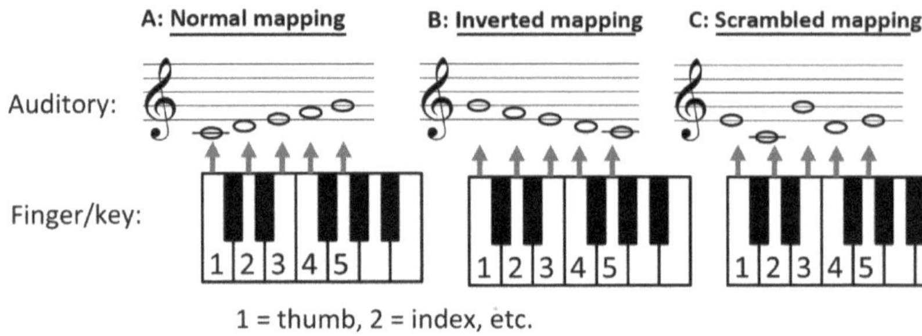

FIGURE 6.3 Examples of how mapping from piano keys (and fingers) to pitch may be manipulated.

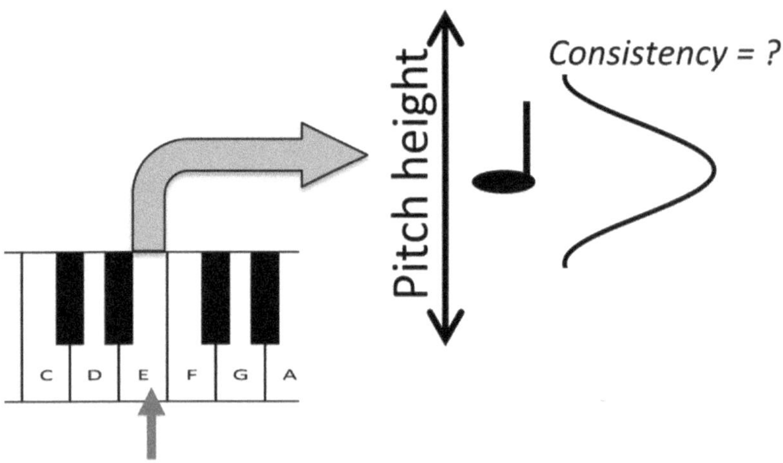

FIGURE 6.4 Reliability constraint on sensorimotor learning.

we press a key and hear a particular pitch class, we anticipate that this same pitch class should result from pressing the same key on future occasions. This constraint can be conceptualized as a transpose of synchrony, as illustrated by Fig. 6.4. While synchrony dealt with consistency in time, reliability deals with consistency in pitch height. In its purest form, this principle would predict that sensorimotor learning might occur successfully for *any* reliable mapping, even those that differ dramatically from the standard way that pitches are laid out on a keyboard. This possibility was tested in a study from Virginia Penhune's research lab at Concordia University (Lega, Stephan, Zatorre, & Penhune, 2016). In one experiment, nonmusicians were trained to associate piano keys to pitches based on a normal mapping of piano keys to pitch, and then play melodies by ear based on that mapping. In another experiment a similar group performed the same task on a keyboard in which the key-to-pitch associations on the keyboard were scrambled, similar to the mapping shown in Fig. 6.3C. Both groups performed the task similarly well—there was no evidence that the scrambled mapping caused a problem for participants in the second experiment. Even though the second group experienced a very odd mapping of actions to sound, the reliability of this mapping allowed sensorimotor learning to occur. However, as we will see, other evidence suggests that the importance of reliability may be limited. Ultimately, reliability may constitute a *necessary* but not a *sufficient* condition for sensorimotor learning.

A major limitation of the first two principles I described is that they are based on *zero-order* associations: that is, associations formed simply between a single planned action and a single outcome. Music comes from relationships across events that come to form a higher-order *Gestalt*, as I discussed in Chapter 4. As such, we must consider principles based on more sophisticated higher-order principles. One such principle is *directionality*: for any change in the direction of movement, a corresponding change in direction should result in perceptual outcome (illustrated in Fig. 6.5). When we first sit down at the piano and press two keys in succession, with the second key to the right of the first key, the resulting change in pitch from one action to the next ascends in pitch. Without any knowledge of the piano's input–output functions we may guess that if we now press two successive keys with movement progressing from right to left, that the resulting pitch change would go down, thus reversing the direction formed by the first two pitches. If the piano did not work in this manner, sensorimotor learning would be made more difficult. Note that it is possible to have a reliable mapping that disrupts directionality; this is exactly what was done in the aforementioned study by Lega and colleagues (2016). It is also possible in principle to have a mapping that causes mapping of actions to pitch that is unreliable yet preserves directionality, although a manipulation of this sort would be more difficult to execute and has not been tested.

Directionality can be considered as an extension of a very basic constraint on sensorimotor interactions. When we reverse the course of any

FIGURE 6.5 Directionality constraint on sensorimotor learning.

direction of movement, the visual field exhibits a corresponding reversal in some dimension. This may be a reversal of optic flow in the case of selfmotion, or may be a reversal in localized motion if we are monitoring movements of our hand while painting. Likewise, when we reverse vocal fold tension from becoming increasingly tense to increasingly relaxed, a reversal in the direction of pitch change should result. Note that this principle makes no claim about the absolute value of a single sensorimotor gesture. Thus, a mirror reversal of visual feedback would preserve directionality, because shifts in the direction of one's movement would still correlate perfectly with directional changes in the visual field. The inverted pitch mapping in Fig. 6.3B also preserves directionality. Likewise, if one shifts from one instrument (e.g., the piano) to an instrument with a different mapping of actions to pitch (e.g., the violin) the directionality constraint would still be preserved. Note that this principle also makes no claim concerning the magnitude of change in perception as related to action. Thus, it is preserved when an individual works with a velocity control system, in which there is an accelerated degree of change in visual feedback relative to a consistent rate of movement. For the present purposes, the most straightforward manifestation of directionality is one's sensitivity to changes in melodic contour in auditory feedback. As such, in light of results discussed in Chapter 4, suggesting a considerable influence of melodic contour on perception/action associations, we may predict that directionality plays a critical role in sensorimotor learning.

Based on evidence to date, I propose that the three principles discussed so far function as the primary task-general constraints on sensorimotor learning in music. Nevertheless, I bring up a fourth principle that likely plays some role in sensorimotor learning, yet seems to influence music learning only weakly. This is the principle of *proportionality*: the magnitude of change in the motor plan from one event to another should be related to the magnitude of change in the perceptual field. Thus, as illustrated in Fig. 6.6, if the first pitch interval on the keyboard that one produces leads to the 9-semitone interval C–A, one might anticipate that a second interval that spans twice that physical

FIGURE 6.6 Proportionality constraint on sensorimotor learning.

distance might lead to a perceived interval of approximately twice that size. Proportionality involves a more sophisticated mapping of action to perception than directionality. Based on experience, we expect that small changes in some kind of motor trajectory will lead to relatively smaller changes in perceptual input than large changes. For example, large movement of the jaw should lead to larger changes in the frequency spectrum of our voice than smaller movements. And large displacements of spatial targets for movements on the keyboard should lead to larger change in pitch than smaller displacements. One possible limitation of this principle is the degree of precision in processing that it assumes. Another, possibly more serious, limitation is that the perception of octave equivalence and tonal relationships in music complicates the conditions under which a pitch interval may be heard as being "twice as large" as another interval.

Over time, these task-general constraints may constrain sensorimotor learning, as mentioned previously. By *constrain*, I mean that these principles set boundaries, or parameters that serve as guides to future learning. There is both a positive and a negative implication of this idea. On the positive side, a constraint can make learning more efficient if a new task abides by the same task-general principles one has acquired in previous skills. In fact, this is often the case, which is why people often find it easier to learn a second musical instrument after learning their first musical instrument. Instruments, as I have argued, tend to share in a common pool of task-general constraints more often than not. On the more negative side, occasionally one has to learn a new skill that does not abide by existing task-general associations and in such cases the constraints can possibly inhibit learning.

Furthermore, task-general and task-specific associations interact and influence each other. Over time, task-specific associations are incorporated into task-general associations as we acquire new tasks. Thus, the distinction between task-specific and task-general associations is only relevant when we compare individuals with different backgrounds. Thus, we can refer to task-specific associations as a possible cause for any differences observed between pianists and nonpianists in the role of auditory feedback during piano performances.

In the following review of existing literature, a few themes arise. First, musical training enhances perception/action associations for action/effect associations for individual pitch events that are specific to a given instrument. Second, task-general associations based on direction and reliability are powerful constraints on perception/action associations, though their roles seem to differ. Third, musical training has a minimal effect, if any, on one's sensitivity to synchronization between perception and action in most cases.

ARE NONMUSICIANS "BLANK SLATES"?

I first consider research concerning to what degree nonmusicians may be said to have no perception/action associations based on sensorimotor associations in the brain, ruling out those based on explicit knowledge such as spatial metaphors for pitch height. Several results suggest that musical training enhances one's sensitivity to specific pitch associations on the keyboard, and on the surface appear to support a kind of blank-slate perspective. In Chapter 4 I discussed how pianists are influenced by the relationship between pitch height and planned movements in simple reaction time tasks (the SPARC effect, Lidji et al., 2007). Comparison data from nonmusicians suggest that their associations may be much weaker, if present at all. Studies of the SPARC effect discussed earlier (Lidji et al., 2007; Rusconi et al., 2006) reported compatibility effects among nonmusicians when pitches were associated with actions toward locations on the vertical plane (up vs down, cf. Douglas & Bilkey, 2007), but not when responses were aligned on the horizontal plane, as on the piano. However, these effects were only present when the task involved categorizing the pitch height, that is, nonpianists were slower if they had to indicate that they heard a "high" pitch by pressing a "low" response key in space. All effects for nonmusicians disappeared when the task involved responding to instrument timbre. Thus, according to this study, the associations found among nonmusicians only appear in tasks that explicitly bring to mind the verbal metaphor associating pitch to space that we use in our language (cf. Eitan & Timmers, 2010). This contrasts sharply with the apparently obligatory associations found for pianists. Similar differences between musicians and nonmusicians were found in the aforementioned study of Drost and colleagues (Drost et al., 2005) concerning visual–audio associations in music.

These results imply that sensitivity to associations between pitch height and motor responses are strong for musicians, so much so that musicians cannot suppress these associations when they impede performance on a task (i.e., the SPARC effect). By contrast, such associations may be much weaker for nonmusicians.[1] Taken together, such associations seem primarily to reflect task-specific associations. It is also significant that the effects discussed so far in this section pertain only for isolated pitches. Significantly, the SPARC effect reduces for both musicians and nonmusicians when responses are based on pitch intervals (Lidji et al., 2007). Nonmusicians can show a SPARC effect when given a contextual pitch that gives them a point of comparison for pitch

[1]The fact that there is some, albeit small, effect for nonmusicians may arise from the explicit knowledge of pitch mapping on the piano that most Westerners exhibit.

judgments (Cho, Bae, & Proctor, 2012), which reduces the need to use task-specific knowledge for the task. In subsequent sections I consider further the significance of tasks using isolated pitches versus pitch patterns.

At the neural level, two studies (detailed in Chapter 2) suggest that training helps form pitch-specific sensorimotor associations. Before any musical training, nonmusicians do not exhibit responses in motor planning areas when listening to melodies, but do show these responses after several days of musical training (Lahav et al., 2007; Mutschler et al., 2007). These results offer strong support for the notion that musical training is responsible for sensorimotor association, and without that kind of specific training such associations may not exist. Moreover, in these studies associations were based on sequential information and not specific to individual pitches. At the same time, even these powerful results face an important limitation. Neither of these studies shed light on whether the nervous system acquires certain kinds of associations more easily than others. The fact that no spontaneous motor activity is elicited by perceptual input doesn't mean that all possible connections are equivalent. As such I next turn to experiments that manipulate the conditions of learning to address more directly whether nonmusicians are truly blank slates.

TASK-GENERAL CONSTRAINTS IN LEARNING

As described above, task-general associations in every case are thought to reflect sensitivity to matching between *patterns of change* in movement to patterns of change in the perceptual field. Perhaps the most basic contingency that an individual may rely on is simple reliability. When all else fails, learning may still occur if a particular action reliably leads to a consistent outcome, even if the pattern formed by planned actions does not have a straightforward relationship to the perceptual pattern that results.

The role of reliability in sensorimotor mapping was addressed in a groundbreaking study by Bangert and Altenmüller (2003). As in later studies (Herholz, Coffey, Pantev, & Zatorre, 2016; Lahav et al., 2007; Mutschler et al., 2007), Bangert and Altenmüller (2003) trained nonpianists to learn simple melodies by ear. Their study was longitudinal, and was carried out across 10 training sessions that took 5 weeks to complete![2] But two other methodological features of Bangert and

[2]This length of training was surpassed only by Herholz and colleagues (2016) who incorporated 6 weeks of training. By contrast, training in Lahav and colleagues (2007) took 5 days, and in Mutschler and colleagues (2007) training only lasted 30 minutes.

Altenmüller were truly critical. First, brain activity in the nonpianists was compared with a group of professional pianists. Second, nonmusicians were divided into two training groups. One group learned melodies by ear in a way similar to other studies. They would hear a short melody, then try to play it back, using the auditory match between feedback from their melody and the target melody they just heard as a gauge of how accurate their performance was. Each correct performance was followed by a more complex target melody, whereas errors led to the repetition of the target melody the participant just attempted to perform. Only five pitches were used in melodies, avoiding the need for participants to modify hand position. A second group completed the same training regime, but experienced uninformative feedback pitches. Each time they pressed a key, a randomly selected pitch from the five possibilities was sampled. The authors called this the "no map" condition based on the fact that participants could not encode any kind of mapping from spatial locations on the keyboard to pitch height. In other words, the reliability principle of sensorimotor learning was disrupted. For this group, training was particularly difficult because they had to determine matches from the target melody to finger movements independent of the pitches that their finger movements might generate. Not surprisingly, this group did not improve in their accuracy of pitch sequencing across the 10 training sessions, as did the other "map" group, although rhythmic accuracy improved with training for both groups.

Bangert and Altenmüller (2003) measured brain activity using a technique one rarely sees in the psychology literature: delayed current electroencephalography (DC-EEG). Most EEG measurements involve rapid fluctuations in the brain's electrical activity as measured on the surface of the scalp, leading to measures like event-related potentials (ERPs) that were discussed in Chapters 1 and 4. By contrast DC-EEG aggregates this activity across very long time frames, 1–3 seconds. This long-term average provides a better estimate of where the electrical activity originates, which is a classic problem in EEG measures given the proliferation of electrical activity throughout the cerebrospinal fluid in the brain. In this sense DC-EEG offers a less expensive alternative to techniques like PET and fMRI, which are optimal for localizing activity, though it is still not as accurate as these measures (see Chapter 1 for further discussion).

The effects of training were measured in two ways. In an "auditory probe" task, participants listened to short pitch sequences on the piano, similar to related studies (Lahav et al., 2007; Mutschler et al., 2007). The second "motor probe" task involved having participants generate spontaneous sequences on the keyboard with the sound turned off. At issue is the degree to which these probe tasks elicited neural activity suggestive of associations between auditory and motor areas of the brain. In comparison to the "no map" group, participants who trained with normal pitch mapping exhibited increased activity in motor areas while hearing pitches. In addition, this group exhibited increased activity in the right inferior frontal area of the brain (possibly the right homologue to Broca's area, see Chapter 4), whether hearing pitches or playing silently on a keyboard. Although it is hard to be certain given their measurements, this brain activity may be based on the right homologue to Broca's area, as was observed in other fMRI studies. These changes in activity led the "map" group's responses to more closely resemble the brain responses of professional pianists over time. In contrast, the "no map" group had more unreliable patterns of activation, despite the fact that they did exhibit learning of rhythm production. Thus, *reliability* of mapping from piano keys to pitch is critical for sensorimotor learning. At the same time, it is not possible to dissociate the role of reliability from *directionality* in this paradigm, because the manipulation of reliability the authors used would also cause directional associations between actions and sounds to vary.

The aforementioned training study by Lega and colleagues (2016) had participants learn melodies based on a mapping that is reliable but does not preserve directionality, and thus offers a useful point of comparison to Bangert and Altenmüller (2003). As noted before, learning did occur in this group; the contrast between this result and that of the "no map" group in Bangert and Altenmüller offers considerable support for the importance of reliability in sensorimotor learning. The two studies complement each other in another respect. Whereas Bangert and Altenmüller analyzed brain activity during learning, Lega and colleagues used repetitive transcranial magnetic stimulation to inhibit activity in the dorsal premotor cortex (see Fig. 4.2), which in turn prevented sensorimotor associations from being formed. These results demonstrate a causal relationship between brain and behavior, and complement other studies suggesting an important role for the premotor cortex in forming sensorimotor associations (Chen, Penhune, & Zatorre, 2008; Chen, Zatorre, & Penhune, 2006).

Further support for the importance of directionality in learning came from another study from Virginia Penhune's lab (Stephan, Brown, Lega, & Penhune, 2016). In this study, nonmusicians were first familiarized with auditory-motor association on a keyboard. Each participant then memorized one of two melodies by listening to it. Following a memory test for the melody, participants engaged in a motor sequencing task in which they had to produce a series of keypresses based on visually presented numeric stimuli (1 = thumb, 2 = index finger, etc.). Importantly, the melody each participant learned could be associated with motor sequencing on a keyboard that was either consistent with the later movement task or incongruent with it, based on directional transitions from one pitch to the next. Participants who first learned a melody that was congruent with the motor sequence did better on the later motor sequencing task than the other group. Thus, directional associations between actions and sound affect the rate of learning even after very brief familiarization with auditory-motor association on a keyboard.

Taken together these studies suggest that reliable and directionally consistent mapping between perception and action facilitates the formation of new perception/action associations in motor planning areas of the brain. Although to my knowledge no studies have independently investigated the role of proportionality, it is important to note that learning in all of these studies occurred whether or not proportionality existed.

TASK-SPECIFIC CONSTRAINTS IN LEARNING

To date most research on sensorimotor learning has focused on learning among nonmusicians. However, the best way to address the effect of task-specific learning is to investigate how well trained musicians can recalibrate their sensorimotor associations when given a sensorimotor learning task that directly conflicts with music-specific associations. A few studies to date have investigated how well pianists can learn to produce melodies on a keyboard with inverted mapping of pitch (as in Fig. 6.3B). As noted previously, an inverted mapping retains all of the task-general associations proposed earlier. Thus, any deficit from performing on such a keyboard should primarily reflect the specific associations a pianist has formed based on experience.

Although several past studies involve impressive longitudinal training designs, the fruits of sensorimotor learning typically appear after even one learning session. In my lab, we took advantage of this fact in a short-term training study that addressed how specific pitch/space associations that are acquired via task-specific associations in musical training may limit one's flexibility in acquiring new sensorimotor associations (Pfordresher & Chow, in press). In this study, pianists and nonmusicians learned melodies by ear, in a training regime modeled after Bangert and Altenmüller (2003). Half the participants learned melodies with a normal mapping of pitch to keys, and this mapping was inverted for the other half of participants, as illustrated in Fig. 6.3B. Note that this inverted mapping retains all task-general principles described earlier. However, we hypothesized that task-specific associations from piano experience may interfere with learning of the inverted map. This is what we found. Although nonpianists were similarly adept at learning melodies based on either a normal or inverted map, pianists were poorer at retaining melodies based on the inverted mapping relative to pianists who learned melodies with the standard mapping. Interestingly, this difficulty among pianists was only found during later recall of melodies. In the initial learning phase, pianists exhibited no trouble learning melodies based on an inverted pitch map.

EFFECTS OF MUSICAL TRAINING ON THE ROLE OF AUDITORY FEEDBACK

Although learning paradigms offer the most powerful way to assess the influence of task-specific and task-general constraints, we can also learn from cross-sectional studies that address how nonmusicians and pianists respond to altered auditory feedback. In such studies differences across groups can shed light on how task-specific associations influence the kind of perception/action associations that are necessary for fluent performance. If an alteration disrupts pianists but does not disrupt nonpianists, that is evidence for the feature that is altered becoming necessary through specific training. By contrast, alterations that similarly disrupt both groups can be taken as evidence for the task-general importance of the feature that is altered.

For evidence of this sort, I reanalyzed data from Pfordresher (2005). In that paper, four different experiments were reported that included a condition in which feedback pitches were selected at random, that is, were unreliably mapped to actions. Two experiments included trained pianists as participants and the other two included untrained participants. The pooled results from these conditions are shown in Fig. 6.7A. In pooling the data we can evaluate contrasting effects across groups with greater statistical power than in the original paper. As can be seen, the effect of randomizing feedback pitch is quite small for nonpianists. The mean of each condition falls within the 95% confidence intervals

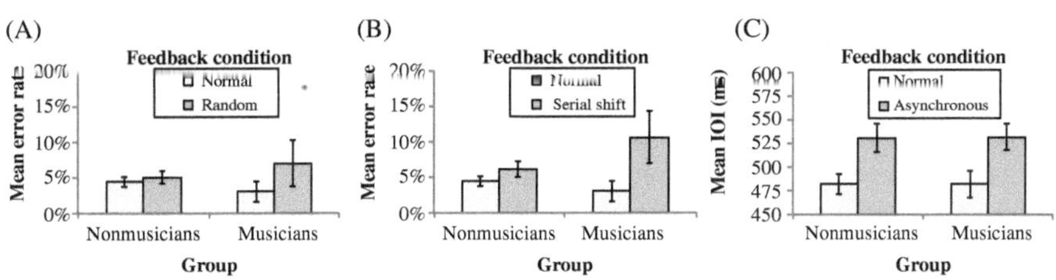

FIGURE 6.7 Effect of musical training on disruption from alteration of feedback pitch content based on randomly selected pitches (A) or a lag-1 serial shift (B), based on pooled data from Pfordresher (2005). By contrast, training has no effect on the effect of feedback synchrony (C), based on pooling data from several other studies. Error bars represent 95% confidence intervals.

surrounding the mean of the other condition. By contrast, there is a stronger effect for pianists, notwithstanding considerable variability in error rates associated with randomly selected feedback pitch.

Here we have a puzzle. Reliability was put forward as an important task-general constraint, and previous studies suggest that unreliable feedback inhibits learning (Bangert & Altenmüller, 2003), whereas learning can occur for extreme alterations to feedback that preserve reliability (Lega et al., 2016). Yet here there was an effect of manipulating reliability only for pianists. The critical difference lies in the role of auditory feedback in each paradigm. In Pfordresher (2005), melodies were learned prior to the presentation of altered feedback. Thus the results do not bear so much on what kind of contingencies are necessary for the formation of perception/action associations as they do on contingencies that must be retained for fluent performance. Because nonmusicians have not formed strong perception/action associations based on specific pitch information, altering these kinds of contingencies may have less of an impact than for pianists. Pianists, for their part, may be more sensitive to specific key–pitch associations because these associations are so well formed. In any case, the lack of an effect of randomizing pitch feedback among nonpianists qualifies the role of reliability in perception/action associations. Reliability may function as a task-general constraint for the purposes of forming associations, but may not be a necessary perception/action contingency if these associations have not already been formed.

Musical training also enhances sensitivity to serial shifts of auditory feedback, as illustrated in Fig. 6.7B. Here, however, there is an important qualification. Although pianists exhibit a larger effect than nonmusicians, there is a statistically reliable effect in both groups. Among nonmusicians, the mean error rate for both conditions falls outside the 95% confidence interval of the other condition. Moreover, every paper that has included nonmusician participants has replicated the statistically significant effect for the lag-1 serial shift for this group. Thus, we can say that the "qualitative" effect of serial shifts (i.e., whether significant or not) holds regardless of musical training, although the "quantitative" effect (magnitude) may vary. This is a critical point because it shows that sensitivity to altered contents is not solely a result of task-specific associations. Task-general features may play a role as well. The effect of serially shifted auditory feedback may reflect the powerful role of directionality in perception/action associations, given the broader literature concerning the role of feedback content discussed in Chapter 4.

In contrast to these effects of feedback content, musical training may have little effect on sensitivity to feedback synchrony in general. Fig. 6.7C presents data pooled across five published studies that all include asynchronies equal to 33% of produced IOIs (Kulpa &

Pfordresher, 2013; Pfordresher, 2003a; Pfordresher & Kulpa, 2011; Pfordresher & Mantell, 2012; Pfordresher et al., 2014). As can be seen, in stark contrast to effects of feedback pitch, the effects of asynchronous feedback on timing is virtually identical across participants with no musical background and those with at least 5 years of piano training.

Regarding synchrony, there is at least one likely exception that to my knowledge still needs empirical verification. Pipe organs often have considerable time delays from key depression to tone onset. Pollard (1968) estimated the delay of pneumatic organs to be between 100 and 150 ms. These delays are long enough to disrupt performance among most performers, as discussed in Chapter 3. Somehow organists must be able to overcome these delays over time, although to my knowledge there are no existing studies comparing organists with pianists on DAF tasks, or even studies that address how much timing precision is present for organ performances. In light of the principles of sensorimotor learning discussed earlier, we can predict that a pipe organ with a variable (i.e., unreliable) delay might elude even the most diligent students.

As noted previously, proportionality is another potential task-general constraint. However, results from studies discussed in Chapter 4 suggest that proportionality plays little role in perception/action associations during music performances. Recall that, for pianists, disruption can occur during piano performances if auditory feedback leads to a serial shift of a contour-matched melodic variation rather than the planned melody. In such cases, the proportionality between the magnitude of spatial displacement on the keyboard to pitch intervals is disrupted, whereas contour is retained. The same kind of result was found for nonpianists as well, thus suggesting that proportionality is a weak associative principle.

To summarize, these data suggest that perception/action synchrony functions as a powerful task-general constraint on performance. Musical training does not alter sensitivity to this contingency. Directionality likewise plays a significant role for both pianists and nonpianists, though pianists via task-specific training may become more sensitive to this contingency. Reliability, on the other hand, only plays a significant role during performances for trained pianists, even though nonpianists depend on reliability during learning.

CONCLUDING REMARKS

In this chapter I have reflected on theoretical principles that may guide the acquisition of sensorimotor associations, focusing on a broad distinction between task-specific and task-general constraints. The role of sound in music performance depends on both factors, with their contributions varying with the kind of experience the performer has. Whereas all performers may be said to share in a similar pool of task-general associations, trained musicians also acquire task-specific associations. These associations lead to constraints on learning, which may make learning more efficient in certain situations (e.g., transfer of training across different melodies on an instrument you have learned), but in other cases may lead to interference (e.g., learning a new instrument with a very different kind of mapping). In the next chapter we consider a kind of behavior for which all of us have highly solidified associations: speech.

CHAPTER

7

Use of Sound in Speech Versus Music

A good friend of mine from graduate school dreamed of someday teaching at a university. Yet he avoided this career completely for one sad reason: he stuttered. Unable to imagine a career delivering lectures that would fall short of the eloquent, dramatic standard to which he aspired, my friend chose another path. And how could anyone blame him? To stutter is to harbor a cruel albatross that hangs over any social interaction, and usually takes its revenge during the most critical, anxiety provoking occasions. To this date, stuttering remains enigmatic to a great extent, despite incredible strides we have made toward understanding and treating it (Howell, 2011a).

I consider stuttering here for two reasons. First, one of the causes of stuttering may be the way in which the brains of those suffering from it process auditory feedback. Second, stuttering often disappears during singing and forms of speech that are preplanned in the way that music often is. Mel Tillis, a famous country-western singer, stuttered so severely during the between-song banter at his concerts that he gave himself the selfdeprecating nickname, "Stutterin' Boy." James Earl Jones and Michael Palin both exemplify mastery of the language as actors in different contexts (drama and comedy, respectively), and yet both suffered from stutters when speaking extemporaneously during childhood. Thus, a sensorimotor affliction that accompanies the production of speech may disappear during the production of music. This apparent *dissociation* (a deficit in one ability that is not apparent in potentially related ability) highlights a deeper question: to what degree are music and language governed by shared or distinct patterns of activity in our brains?

One of the most active areas in music psychology in the early part of the 21st century has involved comparing how the brain processes language versus music. These two forms of communication are often referred to as distinct *domains* of information, which bears on the

DOI: https://doi.org/10.1016/B978-0-12-809196-8.00007-3

question of modularity discussed in Chapter 5. To what degree do our brains treat these domains as independent? There are several reasons why this topic intrigues so many people. Music and speech are similar in many respects, but there are also critical differences between them. Additionally, music and speech abilities are not always consistently related to each other. I detail many of the relevant points below, although the interested reader may wish to consult several in-depth sources for more information (Heffner & Slevc, 2015; Patel, 2008, 2011, 2014; Peretz, Vuvan, Lagrois, & Armony, 2015; Zatorre & Baum, 2012).

At the outset, it is important to clarify that my focus will be on "speech" rather than "language." Whereas language refers to all forms of linguistic communication, including written text and signing, speech refers specifically to the auditory modality and vocal production. Whereas language and music may share similarities at an abstract level, the relationship between sound and action in music is best compared with the relationship between sound and action in speech.

I am primarily concerned with whether perception and action associations found for musical tasks, the main focus of this work, rely on the same mechanisms that are used for speaking. As such, although I will summarize critical research on music and speech in general, I will devote more attention to research specifically oriented toward perception/action associations. With respect to the literature concerning effects of altered auditory feedback on speech, which is extensive, the review here will focus on those studies that form an informative point of comparison with the music studies discussed elsewhere in this book.

COMPARING MUSIC AND SPEECH AS FORMS OF INPUT

Although music and speech both involve the formation of complex, time-varying auditory signals, there are many characteristic differences between the input we receive from each domain. Thus it is possible that the neural resources necessary to optimally process speech versus music may differ. One particularly critical difference between music and speech has to do with the degree of *temporal resolution* versus *spectral resolution* that is required for comprehension (Zatorre, Belin, & Penhune, 2002). Temporal resolution refers to the ability to process rapid changes in a signal over time. In speech, changes in the signal that are associated with phonemes (the smallest meaningful unit of sound) unfold rapidly and thus any system that can comprehend speech must have excellent temporal resolution. For instance, the acoustic signal for consonants in speech is often a rapid fluctuation in frequencies of the voice that takes only a few milliseconds. Spectral resolution refers to the clarity with which the individual frequencies of vibration in the signal can be separated from each other (or "resolved" in the parlance of psychoacoustics). Spectral resolution is critical for music because the constellation of sound frequencies communicates instrument timbre and also helps us perceive pitch (Oxenham, 2012).

Unfortunately a system that exhibits strong temporal resolution typically lacks spectral resolution and vice versa. Cochlear implants, which are used to facilitate auditory perception in the hard of hearing, make an excellent case in point. These sensors were designed to help people comprehend speech, and thus have high temporal resolution but poor spectral resolution. As a result, music perception in cochlear implant users little resembles what the rest of us hear,[1] and people who suffer from hearing loss often do not enjoy listening to music with their implants. Zatorre and colleagues (2002) further suggest that our brains evolved to deal with this tradeoff, and as a result the right auditory cortex exhibits better spectral resolution and is more active during music listening, whereas the left auditory cortex exhibits better temporal resolution and is more active during speech perception.

For the present purpose, this distinction between music and speech is critical regarding the role of sequencing and timing. Speech includes a lot of time-varying information, and speech sounds often carry over into each other. Within this complicated signal, what constitutes a spoken event that is involved in sequential planning? The H-PAC model, proposed in this book, considers events to be rhythmic units. In most languages, including English, the rhythmic units are syllables and so as a first pass I will assume that events in speech are syllables, and that synchronization of perception and action is based on syllable onsets, with sequencing having to do with sequencing of syllables. That being said, it is clear that a great deal of informative time-varying information occurs within a syllable, and this fact is worth bearing in mind as we review the existing literature on the role of auditory feedback in speech.

Musical events are usually considered to be notes, chords, and other temporally isolated sounds whose timing communicates rhythm. The segmentation of musical events is in many cases more clear-cut than for speech, but not always. For instance, singing features dynamic fluctuations in pitch within an event. One kind of fluctuation is the scoop in pitch that occurs at the beginnings of many sung notes, and is used to great effect in the "crooning" style of singers like Bing Crosby. These fluctuations influence listener judgments of intonation and thus add to the content of the event (Larrouy-Maestri & Pfordresher, 2018). In addition, pianists often use rapidly timed ornaments, like grace notes and trills, that are not treated as rhythmic events. Thus, there are

[1]For simulations, visit http://auditoryneuroscience.com/prosthetics/music.

informative fluctuations in music that may not be considered events, though the impact of these changes may not be as extreme as the rapid fluctuations found in speech.

There are also differences in timing that are typically found across music and speech. Music, in the vast majority of cases, communicates a stable referent time interval called the *beat*, and successive beats vary in accent strength to create a meter. As such, timing in music must remain stable enough to communicate this kind of structure. By contrast, speech timing is more irregular, causing many linguists to question whether any kind of higher-order temporal organization exists in speech (Patel, 2008). If speech does have a meter, it might be similar to a complex *heterometric* structure (i.e., more than one meter), which is found less commonly in music (Brown et al., 2017). Furthermore, any kind of metric organization in speech occurs in bouts, with period of metricality being separated by nonmetric pauses.

A byproduct of the metrical characteristic of music is that events repeat more often in music than in speech (Vousden & Brown, 1998). In fact, repetition may be a primary feature of music based on the *speech-to-song illusion* (Deutsch, Henthorn, & Lapidis, 2011). Diana Deutsch, who discovered this and many other auditory illusions, was at home one day listening to a demonstration CD she had recently recorded. Unexpectedly the CD stuck at one part where Deutsch was introducing one of the demos, which caused the phrase "sometimes behave so strangely" to be repeated over and over. When this happened, Deutsch was in another room of the house and thus only somewhat aware of what was happening. Initially she was puzzled to hear singing from the other room when singing was not part of the CD. On further investigation, of course, she discovered that what sounded like singing was in fact repeated speech. This discovery has led to a flurry of related research, including many replications of the original effect (Falk, Rathcke, & Dalla Bella, 2014; Simchy-Gross & Margulis, 2018; Tierney, Dick, Deutsch, & Sereno, 2013; Vanden Bosch der Nederlanden, Hannon, & Snyder, 2015). The effect is highly reliable. The brain even responds differently as soon as participants start to perceive song rather than speech, for the very same acoustic signal (Tierney et al., 2013). Certainly speakers do not attempt to use pitches from musical scales, thus it seems that the periodicity that emerges from repetition is a central distinction between what we call speech and what we call song.

COMPARING MUSIC AND SPEECH IN PRODUCTION

The majority of studies comparing music and speech have to do with perception. However, there are several differences in terms of production that are critically important here and receive less attention in the existing literature.

First, there are differences in the motor systems used. This is even true to some extent in singing. Whereas singing involves using the same general system for producing both verbal (speech) and nonverbal (music) information simultaneously, there is a difference in weighting the importance of muscle groups within the vocal system. Vocal production involves three component systems for the processes of respiration (breathing), phonation (use of laryngeal muscles to control pitch), and articulation (use of the tongue, lips, cheeks, and jaw to shape the sound's frequency spectrum). The musical components of singing rely heavily on phonation. This is the system that allows you to control pitch. Although articulation does affect pitch to some degree, as any good voice teacher will tell you, the role of articulation is primarily to communicate lyrics. More to the point, when reproducing melodies with pitch, it is absolutely critical that you phonate the correct pitches. Any shaping of timbral qualities with the articulators simply builds on that foundation.

By contrast, the communication of verbal information relies critically on articulation. As my daughter was told in chorus, "diction is done with the tip of the tongue and the teeth." Nothing about the larynx in that dictum. Although pitch is important for conveying intentions in speech, its role is not as critical as articulation. A rising pitch can be used to communicate a question, but the habit of "up speech," which is common in younger speakers, involves rising in pitch at the end of a declarative sentence. Moreover, speech can be whispered with little influence on meaning. This is even true of tone languages, for which pitches are associated with word meaning.

I bring this up to make the point that motor learning in the service of speech may not influence music performance skills. A few anecdotes help to clarify this point. Pope Francis, an engaging speaker who is fluent in many languages, is by his own account "tone deaf" and even has trouble with simple chants used in church services. Another talented orator, Hillary Clinton, also suffers from difficulty controlling sung pitch, as was discovered via a chance recording of her attempting to sing the national anthem.

Differences in motor coordination across speech and music performance become more obvious when we broaden our perspective beyond singing. Instrumental performance involves using motor systems in novel ways to control an external device like a keyboard or a saxophone. In so doing, of course, the user not only learns a new motor skill, but also forms new associations from motor planning to perception, as discussed in Chapter 6.

Another important difference in the production of speech and music has to do with the relationship between planning and production. We use speech to convey our thoughts and feelings to other people, who respond to what we say verbally, resulting in an exchange of ideas. In so doing, we most often generate new content as we go along (see Chapter 1 for further discussion). Although talkers occasionally repeat the words, or even the acoustic properties, of other people, such reiterative speech occurs rarely in everyday conversations, being more prevalent in storytelling and comic monologues.

By contrast, musical performances typically reproduce structures that have been composed previously, whether by the individual performing (e.g., singer–songwriter performing her work publicly), or someone else (e.g., a violinist playing in an orchestra). Although performers do add their own nuances to a musical sequence, more often than not the plan of action has been formulated in advance and not spontaneously (an obvious exception being improvisational performance). Thus demands on planning are of a different sort and are most likely less stringent for music. This leads to the difference in the regularity of temporal structure mentioned earlier.

However, the fact that rhythmic organization may not be apparent in spoken output does not mean that plans are not rhythmically organized. Thus, an important experimental concern when comparing music performance with speech is to separate the demands of planning and execution.

THEORIES OF HOW MUSIC AND SPEECH RELATE TO EACH OTHER

The considerable interest in the relationship between music and speech has led to several theories of how these two systems may associate. These theories vary with respect to how much integration there is of music and speech. It is rare to find any theory that proposes either full integration or full separation. This makes good sense. A theory based on two fully separate systems would run into the challenge of explaining how this could happen when music and speech originate in the same auditory signal and are often produced using the same system of effectors. On the other hand, a theory based on full integration would run into the challenge of explaining how a fully integrated system could handle the considerable distinctions between music and language described earlier.

Theories also vary with respect to their level of detail on different components of the underlying architecture. In this spirit I will first discuss a theory that notoriously lacks specifics but has nevertheless been incredibly influential, and is particularly critical for the topic of this book: the *Motor Theory of Speech Perception* (Galantucci, Fowler, & Turvey, 2006; Liberman, Cooper, Shankweiler, & Studdert-Kennedy, 1967; Liberman & Mattingly, 1985). This theory arose from research in the 1950s on speech perception at Haskins labs. At the time, a research team led by Al Liberman hoped to be able to construct a theory of speech perception based on segmenting the speech signal into component meaningful sounds (phonemes), with the ultimate goal of programming machines that could perceive speech. The speech signal was thus conceptualized as analogous to written text, with phonemes mapping onto letters. This notion makes sense insofar as it matches our subjective experience of speech. Unfortunately the speech signal does not match our experience of it. Phonemes bleed into each other—a feature known as *coarticulation*—and the acoustic form of a given phoneme varies from context to context, from person to person, and even just from time to time in general. Confronted with the chaos of the speech signal, accompanied by the near effortless process of perceiving speech, the authors suggested that perceiving speech does not involve perceiving sound! Instead, speech perception may involve decoding the kind of muscle movements, or even motor planning, that lead to the signal.

This proposal on its own was innovative in that it foreshadowed later theoretical developments. For instance, the Inverse Models construct, discussed in Chapter 1, relies on the notion that people can reconstruct a motor plan based on an anticipated perceptual goal. These claims about perception/action associations have remained influential. What differs in the motor theory of speech, and remains controversial, is its restriction of this kind of mechanism to speech perception (Galantucci et al., 2006). Researchers sometimes distill this claim to a simple slogan: "Speech is special." Debates along these lines are referred to as *special mechanisms* debates.

Of course, the motor theory of speech was designed to explain speech perception exclusively. As such it lacks any kind of specification for how we perceive or produce other forms of auditory communication like music, and is likewise silent on the details of speech production, despite relying on production as an explanation for perception.

Later theories offered more specific claims about how both music and speech may be perceived and produced, framed as testable models. Arguably the most well-developed and influential model of music and speech is the modular architecture proposed by Isabelle Peretz and Max Coltheart (2003). A simplification of this model appears in Fig. 7.1. This model was based on accumulated evidence from neurological cases exhibiting dissociations not just between music and speech processing but also between different characteristics of music processing. At the heart of this model is the idea, based on Fodor (1983), that cognitive

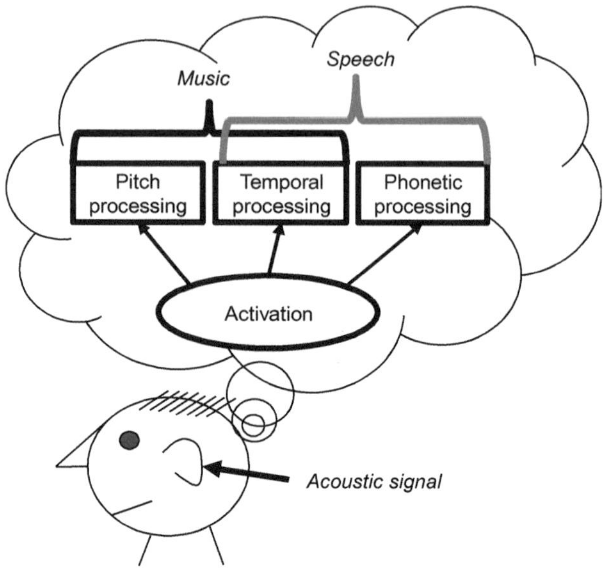

FIGURE 7.1 A highly simplified illustration of music and speech processing that integrates two leading approaches. Different functional modules for music and speech (e.g., Peretz & Coltheart, 2003) are represented by boxes within the thought bubble. The functioning of each module may be governed by a shared resource for activation (e.g., Patel, 2008), shown below the modules.

modules exhibit domain-specific responses. Specifically, each module is tuned to a specific feature of the input it receives. As such, the model proposes a group of modules that respond to pitch but not timing, and within the pitch-oriented modules there is one that responds specifically to the relationship between sequences of pitches to various musical keys, referred to as *tonal encoding* (note that Fig. 7.1 does not represent pitch processing at this level of detail). Perhaps the most critical division within the model is at an initial stage, where the auditory input may be encoded with respect to verbal content (speech) or musical content. In this respect, Peretz and Coltheart's (2003) model may be considered to carry on the special mechanism idea of the motor theory. However, there is more nuance to this model than may be immediately apparent. Although there are several modules that cluster within the music network, some of these respond to the same properties in speech. In particular, temporal organization and the melodic contour of pitch may serve both music and speech.

Although the Peretz and Coltheart (2003) model has had its biggest impact in studies of perception, the model's architecture explicitly describes a process from perception to production, and Peretz (2009) argued that this framework could be used to understand differences between speaking and singing. In particular, the model lays out the chain of processes that may lead to the production of lyrics and tune while singing a melody. As such, it provides a reference point for understanding how auditory information may figure into the production of music or speech. For the most part, this model suggests a compartmentalized use of auditory information across as well as within each domain. However, for certain features of sound, in particular timing, music and speech production may rely on common modules.

A more integrated model of music and language processing was proposed by Ani Patel (2003, 2008). Specifically, Patel proposes that music and language share resources used to decode syntax. This is referred to as the *Shared Syntactic Resource Integration Hypothesis* (or SSIRH). The model therefore proposes a highly restricted version of integration. Integration occurs at the abstract level of syntax: rules for ordering the units of a communication system into meaningful sequences. The model does not make claims about integration for basic features, for instance. Even more important, integration involves sharing of *resources* used to decode syntax or construct a syntactically appropriate sequence. The underlying *representations* that determine syntactic rules are separate. Fig. 7.1 represents this distinction by having the representations based on Peretz and Coltheart activated by a common resource pool.

A later model proposed by Patel also addresses the relationship between music and speech, although it was designed specifically to explain beneficial transfer effects from music training to speech. This is the OPERA hypothesis (Patel, 2011, 2014). The letters in the acronym spell out principles for why learning music may benefit speech processing. Neural resources for speech and music *overlap*, music requires greater *precision* in auditory perception than speech, music engages *emotional* responses, musical practice involves *repetition*, and music is highly *attention*-getting. Although this hypothesis assumes some degree of integration across music and speech, most of the principles leading to this hypothesis focus on distinctions across music and speech. Importantly, the OPERA hypothesis, which succeeded Patel's SSIRH model, proposes a fuller extent of overlap than just syntax, and assumes overlapping resources for basic acoustic features (though possibly not perceptual attributes of those features). This evolution was based in large degree on measures of auditory brainstem responses suggesting that early auditory responses to spoken pitch may be influenced by musical training (e.g., Bidelman & Alain, 2015; Moreno & Bidelman, 2014; Wong, Skoe, Russo, Dees, & Kraus, 2007).

A particularly interesting feature of the OPERA hypothesis is its claim that music processing involves greater precision than the processing of speech (even for tone languages). Patel focuses in particular on pitch precision, thus tying into the idea of spectral precision discussed earlier (cf. Zatorre et al., 2002). A prediction that is consistent with this claim is that sensitivity to pitch in auditory feedback may be greater in the domain of music than in speech.

TIMING AND SEQUENCING IN SPEECH AND MUSIC PERFORMANCE

Bold, empirically falsifiable claims are critical for scientific advancement, and unfortunately not observed as often as we would like. In this respect, Peter Howell's hypothesis articulated some years ago (Howell et al., 1983) will be central to the discussion that follows:

> Consider music played on keyboard instruments. With electronic organs the pitch can be transposed at a flick of a switch without affecting performance. For a musician, it does not matter what the relation between an action and its outcome is; therefore, the feedback is not used to regulate the action. On the other hand, if a musician is required to regulate his or her meter with respect to other players, similar effects to those produced by DAF occur. (p. 773)

This excerpt is noteworthy because it articulates a straightforward claim that comes from a more complex set of theoretical assumptions. First, this statement focuses on the fact that perception/action associations for musical instruments are more flexible than for speech. The specific comment here has to do with changing the key on an organ, but in the broader context one can appreciate the many-to-many mapping of actions to resulting sounds that are possible for instrumental music. On a guitar, moving upwards in pitch can be produced by depressing a string and moving your finger toward yourself along the fret board, which for a right-handed guitarist involves movement to the right. However, on a cello, a similar upward change in pitch may be created by a downward movement. And of course one can change pitch on both instruments by switching from one string to another, which involves moving down on the vertical plane (i.e., in space) on the guitar, or from right to left on the cello. By contrast, changing pitch in the voice usually involves changing the tension of one's vocal folds. Although there are other ways to vary pitch using laryngeal muscles and respiration (for a review, see Sundberg, 1989), it stands to reason that action/perception associations are more constrained in speech.

According to this logic, effects of altering content may differ for the production of speech as opposed to instrumental music performance. The quote given above leaves the nature of this difference unspecified, yet hints at the possibility that may matter more in music than in speech production given the more complex set of associations one has to negotiate for music. In contrast to the effects of altering content, Howell and colleagues predict similar effects for music and speech production when feedback synchrony is altered.

Interestingly, Howell and colleagues' reference to music specifically involves instrumental music with no mention of singing. It is common in music cognition to compare speech to instrumental music rather than singing given that the use of lyrics in song confounds attempts to address music and speech separately. Nevertheless, comparisons between song and speech are useful because differences across domain are not confounded with differences in the motor system. My own read of Howell's argument is that altered feedback should yield similar kinds of effects for the production of song and speech, given that one cannot alter sung feedback like one can with the flick of a switch on an organ (note that Howell's study predates the use of autotune).

Because Howell and colleagues (1983) make explicit associations between music and speech, studies from Howell's lab concerning effects of altered feedback on speech make an excellent point of comparison with previously discussed studies using musical stimuli. As discussed in Chapter 3, Howell's initial work on DAF effects led to the *Displaced Rhythm Hypothesis*, which suggests that the role of auditory feedback is limited to that of temporal interactions between perception and action. Several studies provided strong initial support for this hypothesis (Howell, 1983; Howell & Archer, 1984; Howell et al., 1983). At the time, the prevailing view of why auditory feedback disrupts production had to do with error monitoring: you use feedback to determine if something went amiss between our planning of actions to their execution; a delay in auditory feedback tells you that something went amiss. According to this view, the disruptive effect of DAF happens because of how we respond to these presumed errors. This hypothesis assumes that we interpret auditory feedback as resulting from our actions. So, what happens if one alters the content of auditory feedback so that it could no longer result from our actions? According to the prevailing view, feedback of this sort should not disrupt production, regardless of its temporal relationships to one's actions, because it cannot be interpreted as resulting from an error.

A study by Howell and Archer (1984) provides the best example of how the displaced rhythm hypothesis was tested. In experiment 2 of that paper, participants produced sequences of randomly ordered /i/ and /u/ vowels, which participants read from a typed list as fast as they could without making errors. On a given trial, auditory feedback could be delayed in 100-ms increments up to a maximum of 400 ms. As in other studies, DAF slowed down production, particularly when the delay was 200 ms. The new twist is that on half the trials, the participants' voices were transformed into a square wave tone that had a single pitch and varied only in its loudness, matching the intensity of the participants' voices. This transformed feedback could also be delayed. Herein lies the critical feature: under natural conditions your voice would never sound like a square-wave tone. Moreover, this transformed feedback contains no phonetic information whatsoever and thus would not be processed using speech components in a modularized music/speech system (cf. Peretz & Coltheart, 2003). Therefore, if DAF disrupts performance because it sounds like an error, then delays of feedback that has been transformed in this way should not disrupt production. Yet, Howell and Archer (1984) did find a similar pattern of disruption from DAF regardless of whether feedback reflected the participants' speech or was transformed to a square-wave tone. In fact, delays of the transformed feedback were somewhat more disruptive to the timing of production! Other studies by Howell's group further support the view that DAF disruption is rhythmically based by showing similar disruption when one simply switches feedback off and on in a rhythmically disruptive way (Howell et al., 1983), or when you delay part but not all of the speech signal (Howell, 1983).

Not all of these results match those from music performance tasks. For instance, studies on the effects of DAF on piano performance suggest that DAF effects are reduced, though still significant, when pitch content is randomized (Finney, 1997), which contrasts with Howell and Archer (1984). In general, as discussed in Chapter 4, the very limited role for content proposed in the Displaced Rhythm Hypothesis does not seem to hold for music. Howell (2004a) suggested that these differences may exist in part because of the preplanned nature of music. However, Howell's group did not manipulate content in ways similar to that of studies conducted in music.

In my lab, we have attempted to address the role of content in singing and speaking by using manipulations designed to simulate the effects of lag-1 serially shifted feedback in piano studies (Pruitt & Pfordresher, 2015). We did this by presenting participants with asynchronies that were equivalent to the produced interonset interval determined by a metronome. Fortunately, because serial shifts do not affect timing, we can assume that timing should remain stable, so that each time the participant produces a note or syllable, they hear the last one they produced and feedback will be timed more or less in synchrony with production (or at least not so asynchronous as to cause disruption of timing). The stimulus materials were designed to have the same level of complexity: We generated sequences of nonsense syllables where each syllable mapped onto a pitch used in melodies (C = "loo," D = "bah," etc.). We found that serial shifts (i.e., delays equal to the produced interonset interval) elevated error rates significantly for both sung melodies and spoken syllable sequences. A similar manipulation within speech was reported earlier (Müller, Aschersleben, Esser, & Müsseler, 2000). In that study, participants produced sequences of syllables that varied in the complexity of their metrical properties (syllable stress, e.g., "ta TA") and/or content (e.g., "ta ti") while hearing normal feedback or delays that matched speaking rate. Although DAF slowed timing regardless of sequence complexity, participants only increased their vocal intensity (related to disruption) when content alternated from one syllable to the next.

What do these results mean? As mentioned in Chapter 4, I suspect the difference in results from these to the earlier studies from Howell and colleagues may be because in those studies the altered content did not form a higher-order pattern that conflicted with sequencing. More broadly, these results suggest that the use of auditory feedback may be consistent across music and speech. Of course, in all these studies the speech sequences were not representative of spontaneous speech, which is more complex and makes more demands on planning than repeating strings of syllables. However, the appropriate comparison with spontaneous speech would be improvised music. To my knowledge, that kind of comparison is not yet available.

Pruitt and Pfordresher (2015) also replicated the sequencing/timing dissociation in speech (cf. Chapter 5). Whereas simulated serial shifts, described previously, yielded a larger effect on produced errors than on timing, shorter delays that merely resulted in an asynchrony of produced syllables slowed timing while not increasing errors. However, there was a subtle difference across speaking and singing that reflects an important distinction across domains. Whereas simulated serial shifts had no effect on timing for singing tasks, they yielded a moderate slowing effect for speech. In other words, the dissociation of sequencing and timing was not as "clean" for speaking as it was for singing. The reason likely relates to the problem of units described earlier. There is time-varying information within a syllable. Thus when auditory feedback presents the previously produced syllable as you are speaking, the difference between the planned and actual auditory feedback does not simply reflect categorical content—there will also be temporal differences.

To round out this discussion, Fig. 7.2 presents quantitative comparisons across studies that involve the production of speech or music, where music was produced vocally (like speech) or on a keyboard (using a different effector system). I pooled results from my lab for studies that involved at least two of the three tasks (Pfordresher & Mantell, 2012, which investigated keyboard and singing performance; and Pruitt & Pfordresher, 2015, which compared song and speech), and also incorporated results from two key studies from Peter Howell's lab

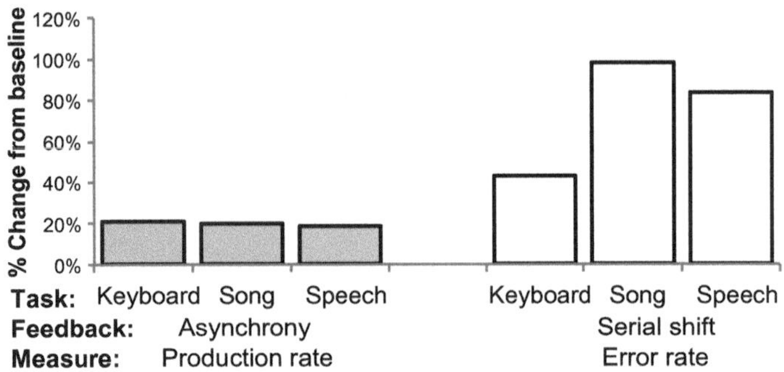

FIGURE 7.2 Comparison of several studies that involve manipulations of feedback synchrony or content, for music and speech production tasks. See the text for further details.

(Howell & Archer, 1984; Howell et al., 1983).[2] Results across studies are standardized by taking the difference in performance across altered and normal conditions, and dividing that difference by the measure during normal feedback conditions, which functions as a baseline. The three bars on the left side represent the effect of asynchronous feedback on the mean rate of performance. As can be seen, these effects are highly uniform across studies. This is true even if you break down the data further by the individual studies that involve speech or singing, which I didn't do for Fig. 7.2 to keep the figure simple. Effects of serial shifts are shown on the right. Note that the Howell studies did not use this manipulation—these data points come from my lab. Effects of serial shifts are noticeably larger for vocal production than for keyboard production. This likely reflects the demands of fine-grained pitch control in vocal production, a topic we turn to next.

FINE-GRAINED CONTROL OF CONTENT IN SPEECH AND SONG

As noted above, a spoken syllable carries meaningful time-varying information, perceived as discrete phonemes. This fact contrasts sharply with music performance on the piano, in which pitches are represented discretely across keys. In all pitched percussion instruments (e.g., piano, marimba), the user exerts no fine-grained control over pitch. In many other instruments, and especially for singing, the performer can

[2]I used data from experiment 1 of Pruitt and Pfordresher (2015). From the Howell studies, I used data from altered feedback experiments on speech that used a delay of 200 ms at the lowest amplification level (which best matches the loudness used in keyboard studies).

modulate pitch, intensity, and timbre throughout the duration of an event, similar to the way in which speakers modulate the spectrum to communicate phonemes. Although such fine-grained control allows the performer more expressive flexibility, it also creates a performance demand not found in the piano.

All of this leads to a different kind of role for feedback content. In Chapter 4, we considered feedback content only at a sequential level. In those studies, for a given produced event, feedback content would match or not match what was intended in a discrete sort of way. Likewise the result of such manipulations for piano performance could be an increase in the production of pitch errors, again considering content as a discrete variable.

An extensive area of research exists concerning how people use auditory feedback in the online control of vocal production. Arguably, the most challenging part of learning to sing is being able to control pitch. Of course, pitch also plays an important role in speech. A kind of feedback alteration is used in most of these studies that involves shifting the frequencies in auditory feedback (the first reported study to my knowledge is Burnett, Kiran, & Larson, 1997). In a typical study, participants will be asked to produce a sustained syllable on a series of trials. On certain trials (often an entire sequence of trials that follow a baseline), feedback from the voice is shifted up or down in pitch, and the researcher measures responses to the shifts.

This paradigm resembles alterations to feedback pitch in piano performance, in that feedback pitch content differs from the anticipated consequence of phonation. However, there are some important differences in this area of research from the kind of manipulations discussed elsewhere in this book. For the most part studies from this area are interested in basic vocal motor control for speech. Related to this observation, these studies typically involve the production of individual pitches as opposed to a melody or a sentence. Thus, issues of melodic organization, which have been so central in studies of feedback content discussed earlier, are not addressed. In addition, the manipulations involved in these studies are generally subtler than those used in other studies. A common shift magnitude is one semitone (100 cents), whereas this alteration is the minimum possible value in piano studies in which the pitch of auditory feedback typically shifts to another category that is minimally 100 cents (one semitone) away. These small shift magnitudes appear to be chosen by design. Jones and Munhall (2000) identify the 100-cent shift as being "intentionally small to avoid detection by subjects" (p. 1249). This stands in stark contrast to the highly noticeable alterations of feedback content used in other studies summarized here. Thus, results from pitch shift studies may involve different mechanisms for associating perception and action than those that are affected by other manipulations of content discussed in Chapter 4.

Two common findings in the literature have important implications for both speech and music. First, responses to pitch shifts—at least early on—appear to be involuntary, leading researchers to refer to such effects as a kind of vocal reflex (e.g., Liu & Larson, 2007). This is particularly true of the first 100 ms of a produced pitch. Even when participants are instructed to suppress their response to a perceived pitch shift (which, as noted before, may be hard to perceive), participants typically shift their produced F0 in the opposite direction of the shift within this timeframe (although responses in the same direction are occasionally found, Franken, Acheson, McQueen, Hagoort, & Eisner, 2018). Such compensatory responses retain the feedback F0 at the desired pitch height.

This effect explains a common problem in choirs: It can be difficult to maintain one's own pitch while singing next to someone who is singing a different part, or perhaps singing the same part as you but with poor intonation. Choir members occasionally mute the sounds of other players by pushing their earlobe into their ear canal, in an effort to overcome these effects. Of course, highly accurate singers do not do this! That's because highly trained singers are better at overcoming the effects of a frequency shift when instructed to do so (Jones & Keough, 2008; Zarate, Woods, & Zatorre, 2010; Zarate & Zatorre, 2008). However, even among highly trained singers there is a tendency to respond reflexively to pitch shifts. A related fascinating finding is that individuals with congenital amusia, a disorder that disrupts awareness of musical pitch and thus melody recognition (a.k.a. "tone deafness") still show this reflexive pitch shift in most cases (Hutchins & Peretz, 2013). So frequency shift effects appear to disrupt the pathway that links perception and action directly, rather than the conscious awareness of pitch (cf. Hutchins & Moreno, 2013; Loui, Alsop, & Schlaug, 2009).

The second critical finding for our purposes is that repeated presentations of frequency-shifted feedback cause participants to adjust their internal model of perception/action associations (see Chapter 1 for more discussion of internal models). The first documentation of this effect was reported by Jones and Munhall (2000). Participants produced vowels with frequency-shifted feedback on a series of trials, and then continued the same task as feedback reverted back to normal. During feedback shifts, participants showed *compensation* for the shift. If the feedback pitch were shifted up 100 cents (a semitone), participants would respond by adjusting their vocal folds to generate what would usually be a lower pitch. Thus, they would execute a motor action normally associated with a musical "E" to maintain feedback pitch that would match a musical "F." Then, when feedback returned to normal, participants exhibited an *after-effect:* rather than reverting back to normal pitch production, participants started singing too high. In other words, they recalibrated pitch production in an extreme manner. Presumably, this overshoot behavior occurred because participants started to hear a shift down in auditory feedback and mistakenly interpreted that as meaning their vocal motor control was undershooting the target to a great degree. This effect is analogous to a classic effect in vision called *prism adaptation,* which was first described by Helmholtz. If you wear goggles that shift the visual field to the left, and are then asked to point to a spot on the wall, you need to learn to adjust motor control so that your reaching movements go further to the right (in the opposite direction) than you would normally use.

A later study suggested that in the singing of melodies this effect could be specific to individual pitches that are shifted within the melody (Keough & Jones, 2011). This brings up the interesting possibility that internal models may not be general to a sensorimotor system but may be specific to planned gestures within that system (Wolpert & Kawato, 1998). Specifically, frequency shifts may cause participants to recalibrate their model of sensorimotor associations, so that when the frequency shift is removed feedback pitch starts to sound too low because the system has come to expect that pitch should be high relative to the motor plan. The adaptation of pitch found in Keough and Jones (2011) is highly relevant for musical contexts, but was inspired by a study that focused specifically on speech (Houde & Jordan, 1998). The paradigms across these studies are similar, but in Houde and Jordan's study, feedback pitch was held constant, while upper frequencies that help to determine the perceived value were altered. As a result when participants repeated the vowel sound "ee," they would hear "eh." Participants would reshape their articulators to keep generating the same speech sound, similar to the adaptation of laryngeal muscles to retain the same pitch during the frequency shift of feedback.

To summarize, research on the use of feedback in vocal production suggests fine-grained control of pitch within spoken or sung events. In contrast to altered feedback effects discussed in earlier chapters, the use of feedback here does seem to reflect a classic closed-loop system, and thus "feedback" in the literal sense. To date there is no suggestion that this process functions differently across music and speech, at least with respect to rapid automatic responses to feedback perturbations. However, acquired singing ability can influence later voluntary responses, which could reflect the incorporation of music-specific processing modules.

A TALE OF THREE MALADIES: SENSORIMOTOR DEFICITS IN MUSIC AND SPEECH

Individual differences in production abilities exist in both music and speech. Whereas most of us speak fluently, a small portion of the population suffers from some kind of speech disorder. Although relatively few people learn a musical instrument, most of the population can sing accurately (i.e., within a semitone of the correct pitch) under the right conditions (Dalla Bella et al., 2007). However, here too a minority of the population exhibits deficient music production abilities. Deficits in production exist in both domains, and in both cases suggest departures from typical development and/or brain function. Here I briefly discuss three distinct deficits that bear on perception/action associations, two of which primarily reflect musical behaviors (congenital amusia, vocal pitch imitation deficit), and the third, which has to do with speech (stuttering). These deficits highlight both the complex interactions between music and speech, as well as the interplay between the use of temporal information and sequential content.

We start with the most well-studied musical deficit: *congenital amusia*, a term that now substitutes for "tone deafness" (for reviews, see Peretz, 2008, 2016; Peretz & Hyde, 2004; Vuvan, Nunes-Silva, & Peretz, 2015). Literally, the term refers to the absence of musical processing that has existed since birth. In practice, congenital amusia refers to an inability to recognize basic musical features, typically related to pitch, without any apparent damage to the brain or deficits in nonmusical aspects of perception, memory, or performance. Diagnosis of congenital amusia comes from an extensive screening procedure centered around a test battery in which individuals have to detect disruptions to unfamiliar melodies based on musical pitch and timing features (Vuvan et al., 2017). Although congenital amusia need not be pitch-specific, most amusic cases are and the vast majority of studies in this area focus on pitch, so much so that congenital amusia is sometimes referred to as a pitch-processing deficit.[3]

Great advances have been made in our understanding of congenital amusia since it was first identified early in the 21st century. To date the leading account places the cause of this deficit *not* on perception/action associations but instead on the ability to form conscious categorical representations based on perceptual input (Hutchins & Moreno, 2013; Peretz & Hyde, 2004). To be clear, congenital amusics typically perform music poorly. In fact, this population often avoids musical activities and so it is hard to find congenital amusics with musical training. However,

[3]More recently, a deficit in tracking the musical beat, called "beat deafness," has emerged (Palmer, Lidji, & Peretz, 2014; Phillips-Silver et al., 2011).

everybody sings. Singing by congenital amusics usually sounds less accurate than singing by the general population (Ayotte, Peretz & Hyde, 2004), and acoustical analyses confirm that congenital amusics sing with more variable accuracy and are more often out-of-tune (Dalla Bella et al., 2009; Hutchins, Gosselin, & Peretz, 2010). Further investigation, however, suggests that the accuracy of congenital amusic singing—though usually poor—in many cases does not show the same deficiency as is seen in perceptual discrimination tasks (Loui, Guenther, Mathys, & Schlaug, 2008; Williamson, Liu, Peryer, Grierson, & Stewart, 2012). In addition, congenital amusics respond to frequency shifts as do typical individuals (Hutchins & Peretz, 2013) and can accurately imitate pitch (Hutchins & Peretz, 2012b; Tremblay-Champoux, Dalla Bella, Phillips-Silver, Lebrun, & Peretz, 2010). Congenital amusia thus may in many cases involve a failure to match two systems, leading to the suggestion that it involves an "action–perception mismatch" (Loui et al., 2009). Occasionally, congenital amusics sing quite well.

Congenital amusia also appears to be a music-specific deficit. Congenital amusics do not show the same deficits in many tasks involving speech recognition as they do in music recognition (e.g., Ayotte, Peretz, & Hyde, 2002). Somewhat puzzlingly, amusics are able to process linguistic messages that are dependent on pitch, such as discriminating a question from a statement. This dissociation may be based on stimulus properties. Pitch changes in speech that convey focus ("They are FRYING chickens," vs "They are frying CHICKENS," Speer, Crowder, & Thomas, 1993), or convey whether the speaker is asking a question or making a statement, tend to involve large changes in pitch relative to music. Moreover, as I have noted before, music perception may require greater precision of pitch perception in general than speech, and this precision may be lacking for those with congenital amusia.

In this respect, congenital amusia may be the converse of a related deficit: vocal-pitch imitation deficit (VPID), otherwise known as poor-pitch singing (Berkowska & Dalla Bella, 2009, 2013; Dalla Bella & Berkowska, 2009; Pfordresher & Larrouy-Maestri, 2015). Unlike congenital amusia, VPID is classified based on the inability to match pitch while singing. Although a small number of VPID cases also exhibit pitch perception deficits in most cases perception appears to remain intact (Pfordresher & Brown, 2007). Thus, VPID may involve another kind of perception/action mismatch: the ability to consciously process pitch along with an inability to imitate pitch. Specifically, this deficit may involve the failure to associate an auditory image of the planned melody with an appropriate motor plan (Pfordresher et al., 2015), due to the complexity of negotiating this process within the vocal–motor system (Hutchins & Peretz, 2012a).

Further support for a dissociation between congenital amusia and VPID comes from responses to altered pitch feedback. A related study of the singing deficit VPID addressed how this group responds to serial shifts of feedback pitch content while singing melodies (Pfordresher & Beasley, 2014). Although VPID individuals produce pitches inaccurately, they are more accurate at reproducing melodic contour, often doing this as accurately as the general population (Pfordresher & Brown, 2007). Thus it is possible to detect disruption from serially shifted feedback during VPID singing by analyzing errors of melodic contour. Pfordresher and Beasley (2014) found that participants generating more extreme pitch errors on average exhibit a smaller response to serial shifts. This result converges with the theory that VPID arises from insufficient internal models of the auditory–vocal system (Belyk, Pfordresher, Liotti, & Brown, 2016; Pfordresher & Mantell, 2014; Pfordresher et al., 2015). In this context, the fact that this response was not found for all amusics further suggests that some amusics in the sample of Hutchins and Peretz (2013) may have also suffered from the production deficit VPID.

VPID also carries over to speech, unlike congenital amusia. Although people generally match sung pitch more accurately than the pitch associated with speech prosody, poor imitators in one domain generally imitate pitch poorly in the other domain as well. This may reflect greater integration of perception and action for vocal production than in perception. Whereas experience learning a tone language, in which pitch conveys lexical information, has fairly tepid effects on pitch perception (Bidelman, Gandour, & Krishnan, 2011; Bidelman, Hutka, & Moreno, 2013; Pfordresher & Brown, 2009), a larger advantage in vocal-pitch imitation is found (Pfordresher & Brown, 2007).

Which brings me to the third malady: stuttering. This speech disorder typically appears in childhood, more often among boys than girls, and often—but not always—disappears with time. The two primary symptoms of stuttering are prolongations (drawling out or repeating a syllable) and hesitations (getting stuck before producing a speech sound). As noted at the beginning of this chapter, one of the mysteries of stuttering is that it often disappears when people either sing or act out a part. It also diminishes when auditory feedback is altered or masked. This was shown to great dramatic effect in *The King's Speech*, a 2010 film that dramatized the difficulty King George VI had as a stutterer when he needed to act as a voice of hope and strength for England on the brink of World War II. One of the more accurate portrayals in the film is a scene in which the king speaks while hearing loud music that masks the sound of his own voice (Howell, 2011b). It has been known for some time that various alterations to auditory feedback can temporarily alleviate stuttering (for a review, see Howell, 2011a).

These effects are fascinating and, at first glance, paradoxical. After all, most of this book has considered how altered feedback *disrupts* production of music and, in this chapter, speech production among typically fluent speakers. This brings up a compelling question: Does stuttering result from some kind of abnormal mapping between perception and action? If so, do stuttering effects also carry over into musical behaviors, as seems to happen in vocal imitation deficits?

As you may expect, many theories of how stuttering occurs and may be treated have emerged over the years. Of those theories that concern perception/action relationships a rough division can be made between those that propose a kind of sensorimotor deficit in stuttering, and those proposing that feedback manipulations can stabilize what is effectively a deficit in speech planning. On the former side, a group led by Joseph Kalinowski at East Carolina University argues that stuttering may result from insufficient engagement of the mirror neuron system during speech (Kalinowski & Saltuklaroglu, 2003). Feedback alterations, according to this view, may reengage this system and thus benefit speaking. This group particularly emphasizes the utility of "chorused speaking," which is when an individual speaks in synchrony with others. This occurs during prayers at church services, and is similar to music-making in its use of contagious group synchronization (cf. Brown, 2007). A related model invokes the use of internal models rather than mirror neurons (Max et al., 2004). By this view, stutters have an impoverished representation of the sensorimotor speech system. As a result, they have trouble both forming motor plans based on the speech goal (inverse modeling), and also have trouble using auditory feedback to correct their utterances (forward modeling). Altered feedback may facilitate stuttering because it activates the auditory cortex and thus facilitates modeling of the target output. What these two perspectives have in common is that they both suggest that stuttering may result from some deficit in relating perception to action in stuttering, which can be ameliorated by alterations of auditory feedback. Both of these accounts receive some support from neuroimaging results suggesting reduced activation of the auditory cortex and increased activation in speech motor areas among stuttering individuals compared with fluent talkers (Brown, Ingham, Ingham, Laird, & Fox, 2005).

An alternative perspective comes from the EXPLAN model described in Chapter 1. According to EXPLAN, stuttering results from a failure to coordinate speech planning with motor execution, and not from a sensorimotor deficit. According to this theory, stutterers attempt to speak before speech plans are fully formed. The various forms of disruption that stuttering takes manifest in the ways that the motor system may respond to a poorly formed plan. One may halt until the plan is better formed, or may perseverate on a given sound, for instance. Altered feedback reduces the symptoms of stuttering because it slows down the timing of motor movements, which causes execution to synchronize more effectively with planning. Note that in this account there is no problem of sensorimotor integration, even though interactions between perception and action can be used to alleviate symptoms. With respect to neuroimaging data, this account points to activity in the cerebellum as particularly informative. The cerebellum has been associated with timing of rapid movements (Ivry & Spencer, 2004), and activation in this area may reflect the kind of timekeeping problems that could be faced by stutterers.

Note that these different accounts situate stuttering at different levels identified by the H-PAC model. Whereas accounts based on mirror neurons and internal modeling suggest a problem of integrating the sequential content of action with planned outcomes, EXPLAN focuses on coordination of perception and action at the level of temporal synchrony.

BENEFICIAL TRANSFER BETWEEN MUSIC AND SPEECH

Interactions between music and speech can have facilitatory effects and even be used to treat certain sensorimotor deficits. I have discussed examples of beneficial transfer in this chapter already. Here I highlight an important clinical intervention known as *melodic intonation therapy*, used to treat disruption of speech after a stroke (Schlaug, Marchina, & Norton, 2008) or in developmental disorders like autism (Wan et al., 2011). This form of therapy harnesses the music-like features of language. In the early stages, the therapist sings words and phrases to the client, while tapping the client's hand in synchrony with each syllable. The client then imitates the therapist as best they can. Over time the therapist models more speech-like forms of communication, until eventually the client can produce speech much more fluently. There have been some impressive successes with this treatment, including the rehabilitation of speech in former U.S. congresswoman Gabrielle Giffords following a severe gunshot wound (Moisse, Woodruff, & Zak, 2011).

There are several interesting features of melodic intonation therapy. First, it uses the same call-and-response procedure that singing instructors use, and the client imitates the therapist just like a singer attempts to imitate a melodic pattern that they hear. Second, the incorporation of tactile as well as auditory stimulation creates a multimodal experience for the patient that highlights rhythmic information.

At present, scientists believe that this therapy works by capitalizing on both the commonalities and distinctions between music and language (Norton, Zipse, Marchina, & Schlaug, 2009). For instance, expressive aphasia (i.e., an inability to speak) that is caused by a stroke results from loss of tissue in the centers of the left hemisphere that contribute to speech production (e.g., Broca's area). The music-like patterns used in the early stages of melodic intonation therapy are thought to stimulate areas in the right hemisphere typically used for processing musical information. In this sense, one "tricks" the brain into using music-related areas for speech, and over time neural pathways (hopefully) reorganize so that previous "music" networks become "language" networks. Evidence for this effect was found in an 11-year-old stroke victim, who regained speech following this therapy and exhibited enhanced activations in brain regions in the right hemisphere (Zipse, Norton, Marchina, & Schlaug, 2012).

CONCLUDING REMARKS

Although this book primarily concerns music performance, there are natural points of overlap with speech in terms of how music is conveyed (via sound) and its acoustic structure. Moreover, people mostly listen to songs rather than instrumental music, and thus derive pleasure from the interplay between verbal and nonverbal information. In this chapter I considered the complex relationship between music and speech as it is presently understood, and how this relationship bears on the way people integrate sound and action during music performance or speech.

CHAPTER

8

Auditory Feedback and Higher Cognitive Functions

At the end of the theme song that initiated each episode of *The Muppet Show*, the character "Gonzo the Great" would descend toward the stage with a horn to blare the ending note. A different sound came out each week, much to Gonzo's surprise and dismay. After each of these disappointing finales, Gonzo would gaze dejectedly into the bell of his horn. He knew it came out wrong. And so did all the viewers, though we experienced a chuckle rather than dismay.

The awareness Gonzo exhibited—of knowing what sound should come out, knowing that the *wrong* sound came out, and wondering why—is what this chapter concerns. It is important to point out that awareness is not necessary to account for the effects I have discussed in other chapters. Indeed, the H-PAC model I have proposed (see Chapters 1 and 5) rests on the assumption that the relationship between perception and action in music is automatic and implicit. Effects of altering auditory feedback are therefore not easily overcome by conscious effort, and the use of conscious executive processes in monitoring auditory feedback may not arise as a byproduct of self perception but instead may involve additional demands.

At the same time, we have all had experiences like Gonzo, and the occasions on which auditory feedback defies our expectations can be salient and can have important consequences. It would be shortsighted to deny the significance of how perception/action associations interact with *higher-order processes*. By higher-order I mean cognitive functions that are thought to serve a broad range of situations and that operate at an abstract level. In cognitive psychology, higher cognitive processes integrate sensory information and perform operations on that information in an amodal (rather than modality-specific) form. Working memory, for instance, involves a "central executive" function that prioritizes how attention consciously accesses information from different modalities

Sound and Action in Music Performance
DOI: https://doi.org/10.1016/B978-0-12-809196-8.00008-5

(auditory and visual). Categorization, another higher process, involves relating specific sensory input to one of a whole battery of types that the present input may represent. In this chapter I consider several ways in which the role of sound in music performance interacts with these higher processes.

THE EXPERIENCE OF AGENCY AND AUTHORSHIP

Agency refers to "the impression that people willfully control their own actions" (Repp & Knöblich, 2007, p. 469). It is a form of *metacognition*, referring in general to times when you reflect on your own state of mind (e.g., "Do I know the answer to this test question?"). One aspect of agency is the experience that the perceived consequence of our actions do in fact reflect what we intended to do: an experience sometimes referred to as *authorship* (Wegner, Sparrow, & Winerman, 2004). Most of the time we take authorship for granted, but this experience can be disrupted, for instance, during states in which we lose motor coordination, and in schizophrenia. This aspect of agency matters most for understanding the relationship between perception and action planning in music.

Agency plays a critical role in many theories of auditory feedback. Consider the classic closed-loop feedback control theory that drove early theorizing about the role of auditory feedback (see Chapter 1). These theories assume that the performer interprets auditory feedback as an error, and that disruptive effects of altered feedback occur because of (mistaken) attempts to correct these perceived errors. For any of this to hold, one must assume that the performer experiences authorship. If you know that the sound that accompanies an action has nothing to do with that action, how could your perception of the sound be used to gauge the accuracy of that action? Of course, in many cases the performer may feel uncertain. This is the humor exhibited in *The Muppet Show*. When Gonzo stares down the bell of his instrument, we know he is wondering whether he in fact was responsible for the unexpected auditory event.

Under what circumstances do we experience agency and authorship? A series of experiments conducted in Japan revealed a close correspondence between temporal synchrony and agency (Sato & Yasuda, 2005). Based on claims by the social psychologist Daniel Wegner (Wegner et al., 2004; Wegner & Wheatley, 1999), these researchers hypothesized that perceived authorship occurs when the timing and content of feedback is plausible within the individual's forward model (see Chapter 1). First, participants experience a learning phase in which they are trained to associate pitch height with the spatial location of keys on a computer, similar to how a piano is configured (right = "high," left = "low"). Then on later trials participants would press a button on the right or left side of the computer, followed by auditory feedback that could differ from the pitch height relationships used in training. In addition, the auditory feedback may also be delayed in time. Both feedback alterations (synchrony and pitch content) reduced the sense of agency relative to cases where feedback matched the participant's expectation. More surprising, it did not make a difference whether sounds actually resulted from participants' actions or were generated by a computer algorithm that simulated the typical timing of participants' responses. The experience of agency was also not reduced in trials when participants made errors, which suggests that people have some advance understanding of whether they are about to produce an erroneous response. In other words, one's forward model may be able to predict an outcome that is unintentionally wrong as well as an outcome that is intended. A subsequent study further suggested that the experience of agency may be more strongly driven by these basic sensorimotor associations, in contrast with our conscious knowledge about whether our actions actually lead to the percepts we experience, as opposed to some external agent like a computer (Sato, 2009).

Our experience of agency may therefore, disconcertingly, be more malleable than one would expect. In fact, experiments have demonstrated surprising illusions of self-agency. One such illusion was documented in speech production. In contrast to studies summarized so far, in which feedback conditions varied across trials, these authors addressed the consequences of persistently hearing the same kind of alteration (Zheng, MacDonald, Munhall, & Johnsrude, 2011). On each trial, participants were cued to produce either "day" or "too." Across a series of trials, they would hear these same words in auditory feedback; however, the voice would be of another talker and at a different vocal pitch than their own. Over time, participants increasingly reported experiencing the altered voice as being an altered version of their own voice, rather than somebody else's voice. That happened even though participants consciously knew it was not "really" their voice. In addition, participants adapted their own vocal pitch in a way that converted toward the pitch of the altered voice, as if they were trying to make their voice "fit" the sound that strangely started to feel as if it belonged to them. Importantly, this assimilation in participants' responses is exactly the opposite effect to what is found in studies where one's own vocal feedback pitch is perturbed (see Chapter 7), suggesting that participants who experienced the illusion of self-agency were *not* experiencing the altered voices as if they were perturbations of self-feedback.

This phenomenon in speech was inspired by a fascinating phenomenon known as the *rubber-hand* illusion (Botvinick & Cohen, 1998). A participant sits at a table and rests one hand on the table behind a vertical board that blocks any vision of the hand. The experimenter places a rubber hand right next to the participants' hand, and in front of the board so the participant can see it. This rubber hand is oriented so that it could potentially be where the participant has placed her hand, although clearly the participant knows that her hand is behind the board. Now the fun starts. The experimenter simultaneously strokes both the participant's hand and the rubber hand with feathers. The participant can feel the stroke on her own hand (which she cannot see), and can see that the experimenter is at the same time stroking a rubber hand that is not hers. However, participants begin to have a curious experience: that the tactile sensation from the feather actually belongs to the *rubber* hand, not her own (invisible) hand. In other words, our participant begins to get the sense that she can actually "feel" sensations from the rubber hand. . .that the *rubber hand belongs to her.*

Taken together, these results suggest that agency, clearly a high-level metacognitive process, can in some situations be determined by low-level sensorimotor congruency. Thus, it is likely that auditory feedback alterations in music influence the sense of agency for performers. Beyond this a deeper, more important question emerges: To what extent is the disruptive effect of feedback alterations related to possible disruption to one's experience of agency?

My lab addressed this issue with the help of then-graduate student Justin Couchman, who studies metacognition in humans and monkeys (Couchman, Beasley, & Pfordresher, 2013). Participants, who were mostly nonpianists, performed melodies from memory while hearing feedback that could be altered with respect to its synchronization and/or content. Furthermore, we manipulated each type of feedback in multiple ways to vary the similarity of auditory feedback patterning relative to action planning. Alterations of content could involve serial shifts of pitch, which entirely preserve sequence structure and only alter its serial offset relative to planning, or randomly selected pitches. Similarly, we altered synchrony in a predictable way (a consistent proportion of mean IOI), or by using a delay amount that was randomly selected to fall in between 100 and 400 ms. These randomized conditions should maximally disrupt agency because they are inconsistent with any expectations participants may have, based on conscious expectations or sensorimotor congruence (cf. principles of sensorimotor learning in Chapter 6). On each trial, participants experienced some combination of these feedback manipulations, or normal feedback, and then participants were asked to rate "how strongly you feel like the sound was produced or not produced by you."

Ratings of agency in Couchman et al. (2013) converged with Sato and Yasuda (2005): Agency ratings diminished as the feedback pattern became more dissimilar to the planned action sequence. A serial shift lessened agency ratings from 100% certainty that they produced the sound (found in trials with normal feedback) to ratings of approximately 70%. Randomly selected feedback pitch further lowered agency ratings to around 55%. A similar pattern was found for alterations of synchrony, with lower ratings for randomly selected delays than a consistent delay. Thus, with respect to agency ratings, the main contribution of Couchman et al. was to address the role of perception/action congruence at the level of sequences.

How did these agency ratings correlate with the disruptive effect of the same feedback alterations on performance? We already know from Chapter 4 that increasing the dissimilarity of a feedback melody to the planned melody does not lead to a linear increase in disruption. The effect is nonlinear. Similarly, although randomly selected pitches caused the lowest agency ratings in Couchman et al., this same condition did not lead to the highest disruption. A similar pattern was found for alterations to feedback timing: in general, the most disruptive feedback manipulations are those that lead to intermediate agency ratings. Disruption is therefore associated with uncertain agency. The most disruptive kind of auditory feedback is not so dissimilar that you can rule it out as unrelated to your action plan, while at the same time it does not quite seem like what you planned to do!

Is there a pattern of causation here? In other words, does disruption from altered feedback happen *because* we are not sure whether we are authors of auditory feedback, or conversely do we become uncertain about authorship *because* we have so much trouble with the present relationship between action and perception? Results from Couchman et al. suggested that neither hypothesis holds. Instead the experience of agency runs in parallel with the effect of altered feedback on action planning. In the original paper, this conclusion arose from a statistical test that determines whether one predictor variable *mediates* (i.e., exerts an influence over) the effect of another variable. However, a simpler result can suffice as well: Effects of pitch alterations on agency are exactly the same when looking at trials that show no disruptive effects of feedback alterations (i.e., error-free trials with stable timing), which is also what Sato and Yasuda (2005) found.

These results suggest that perception/action associations do influence metacognitive aspects of music performance. However, their influence on metacognition may be based on a different system than the kind of system hypothesized in the H-PAC model.

One limitation of the studies discussed so far is that they rely on self report. Any time a participant is asked to reflect consciously on his or her state a number of other mental processes come along for the ride, and these may not be of interest to the researcher. The participant may start to second-guess the intentions of the researcher and what that person may want the participant to do (these are called *demand characteristics*), for instance. Based on these problems, more powerful paradigms can arise if one can identify a behavior that measures the kind of response that is of interest that does not require conscious self report.

One such paradigm is *pseudosynchronization*. Here is the logic. When people tap in synchrony with a regularly timed metronome, they routinely time their taps a few milliseconds before the metronome onset. Participants are not aware of this anticipatory behavior because the anticipation is quite brief (about 10 ms). Nevertheless, participants in these tasks *are* aware that the metronome sounds do not result from their actions. That is, their tendency to anticipate the metronome may reflect their sense of nonauthorship for the metronome tones.

So, what happens when in the middle of one of these synchronization tasks, the generation of sounds suddenly switches, so that the sounds no longer come from an independently timed metronome and instead come from the participant's actions (tapping)? If you still think you actions author the sounds you hear, then you would continue (subconsciously) attempting to position taps slightly before the tone onsets, as if they were from the metronome. The result, thus, should be that you gradually speed up in your rate of tapping, which is exactly what was found when Paul Fraisse, an innovative French psychologist, introduced this paradigm (Fraisse & Vouillaume, 1971). Pseudosynchronization refers to this tendency to speed up, as if the participant is trying to synchronize to herself. The effect even happens if the participant knows that there will be a transition in the origin of the sounds, although the tendency to speed up is reduced by this foreknowledge. Knowing when to expect the transition likewise does not alter the effect (Flach, 2005).

Pseudosynchronization also fits into the general theme so far that the experience of agency, although an aspect of high-level metacognition, relies on fairly low-level sensorimotor processes. This point was explored by Bruno Repp, a leading researcher on the cognitive bases of music performance and timing, who joined forces with Gunther Knöblich, a specialist in the social cognitive aspects of perception/action associations (Repp & Knöblich, 2007). They incorporated a self report measure to determine the role of agency in the pseudosynchronization effect. During trials similar to what I described earlier, participants were asked to identify the point where they felt like the tones started resulting from their own actions rather than a metronome. Critically, the authors compared trials in which participants tapped with other trials in which participants listened to recordings of sound patterns from other performers. In the latter sort of trial, a participant would have to rely on perceiving the variability of timing, which would be greater when the source of the sound switched from metronome to human tapper, and would not have any awareness resulting from the relationship of sounds to planned actions. Detection was much more accurate when participants produced taps, rather than just listened to them, which illustrates the importance of action planning in the experience of agency. During trials in which participants actively tapped, changes in the experience of agency were associated with deviations in produced timing that may serve as cues for agency. In other words, if you find that tones synchronize with your finger taps even when you are a little late, it's pretty clear that the sound no longer comes from a metronome!

Thus, perception and action have a strong relationship to the experience of self-agency, and in particular one's sense of "authorship" over the auditory consequences of actions. However, agency does not directly affect sensorimotor experience, but instead may arise as a byproduct of our sensorimotor experiences.

At the same time, individual differences in the experience of agency may be related to how that individual responds to feedback alterations. In a study that included perturbations of vocal pitch, Subramiam and colleagues (2018) found that individuals who are more accurate at judging self versus external control in a separate "reality monitoring" task also exhibited greater precision in adaptation to feedback alterations. In this case, self-agency judgments were based on a task that measured how accurately participants remembered whether they or the experimenter completed the last word of a sentence. After seeing this result, I returned to the data of Couchman and colleagues and found that a roughly analogous, though nonsignificant, result arose there. Participants who generally reported higher self-agency in conditions for which the feedback was somehow related to production were in general less disrupted by those feedback alterations. Thus, in music as well as speech, people who are better at assessing authorship over the consequences of their actions may be better at adapting to situations where those consequences diverge from their expectations.

ERROR MONITORING

An important process that may be contingent on the experience of agency is our ability to monitor feedback for produced errors. If I sing in a choir next to a performer who produces the melody incorrectly, I can generally rule out the wrong notes as being "my fault." I do not attribute authorship to those sounds, and so although they may affect my ability to coordinate perception and action I do not consider these sounds to result from errors on my part. Our ability to keep track of whether the results of our actions accord with our intentions is called *error monitoring*. Error monitoring is critical for music performance, but maybe not as critical as one might think. Once you get in front of an audience at a concert, error monitoring can destroy your performance. A critical trick every performing musician learns is that the audience typically won't notice your errors, but they sure will notice any attempt to stop, go back, and correct an error once it's been made!

Several studies that I have reviewed in previous chapters suggest that disruption from altered feedback does not happen because participants mistake altered feedback for an error they made. Granting this interpretation, it is still possible that people may—from time to time—make this sort of attribution error.

At the outset, it is important to note that error monitoring functions differently in music performance than in other domains. When we speak, it is important to correct an error once it is made, because the main goal of speech is to communicate concepts to the listener, that is, to sequence content correctly. As a result, speakers will stop and recorrect a speech error as soon as they can, and in so doing will repeat only as much of their speech as is necessary for comprehension (Levelt, 1983). In other words, for most speaking the maintenance of a stable rhythm matters much less than sequencing (see Chapter 7). Similar rapid correction is important in most other domains: consider the role of error monitoring when driving a car, for instance! In fact, people often seem to have a sense that an error is "coming" before it even occurs. Based on these results, Levelt (1993) suggested that error monitoring in speech may be partly internal—based on a prediction of what you are about to produce—as well as being based on auditory feedback.

Music is different from speech, as noted previously, in that maintaining a regular rhythm can be more important than correcting errors of sequencing. Related to this, one of the least effective practice strategies involves stopping as soon as one makes an error and restarting the piece (Mishra, 2002). Whereas beginning pianists usually make many errors of rhythm (stopping, restarting, etc.), expert pianists show the opposite tendency (Drake & Palmer, 2000). That being said, pianists may not be entirely able to suppress the tendency to recorrect the errors they make. Pianists play errors more hesitantly—with a slightly delayed onset and quieter intensity—as if they are trying to withhold producing them (Repp, 1996). Yet they still *play* these wrong notes and—most important—don't go back to recorrect them. Thus, although error monitoring is undoubtedly important for music performance, the degree of importance and application may differ considerably from other domains.

In general, the brain may suppress perceptual input that coheres with our action plans. For visual input, this leads to a kind of "blindness" for response–compatible stimuli. For instance, if a participant plans a rightward action in response to a cue (an arrow pointing right), that individual will find it harder to detect an object presented to the right (in the direction of the planned action) than the left (Müsseler & Hommel, 1997a, 1997b; Stoet & Hommel, 1999). A similar kind of response happens in music performance: Performers find it easier to detect an alteration to an isolated feedback event and find it harder to confirm that feedback matches their action plan—a sort of "deafness" effect to complement the result from vision (Pfordresher, 2014).

These suppression effects likely reflect the role of forward internal models, discussed elsewhere in this book (see Chapter 1). A detailed computational hypothesis of how this suppression works was articulated in the state feedback control model of Houde and Nagarajan (2011), an internal model for sensorimotor interactions in speech. According to this model, the forward model's prediction and input from auditory feedback cancel each other when they match, whereas nonzero values indicate some kind of error. Thus this model offers a computational framework for suppression of outcomes that match one's intentions. Suppression can also be observed at the neural level, with the brain's early response to change in auditory input, the N1 ERP component, being suppressed when those changes arise from auditory feedback during a piano performance (Bendixen, SanMiguel, & Schroger, 2012). Altered auditory feedback may influence these responses. In Chapter 4, I discussed a study by Mathais and colleagues (2017), in which the authors found that alterations to feedback content during piano performance may reduce this suppression, although their results were specific to alterations that present the pitch of the very next future event. Altered feedback events, unlike events that we plan to produce (i.e., correctly sequenced events) thus may take on greater salience by virtue of their discrepancy from one's forward model.

Does this mean that they are interpreted as errors? Pfordresher and Beasley (2014) addressed this question behaviorally, simply by asking participants to estimate their error frequencies at the end of each trial. As described in Chapter 4, feedback alterations in this study were present on a randomly selected subset of sequence events at varying frequencies. Feedback manipulations thus simulated the effect of making a varying number of pitch errors on each trial, although the frequencies used for feedback alterations were overall higher than error rates that pianists and nonpianists typically generate (usually around 5% of all sequence events). Participants significantly overestimate the number of errors they made in all altered feedback conditions, though not during normal feedback. Furthermore, in a pilot study that used random-pitched feedback, this overestimation tendency did not emerge. Thus, altered feedback is potentially confusable with the errors one actually makes. At the same time, estimated errors were far closer to the errors actually produced than they were to the frequency of altered events.

EXECUTIVE CONTROL

An important point about the higher cognitive functions discussed here is that they require *attention*. By contrast, sensorimotor associations may rely entirely on automatized and implicit (i.e., nonconscious) processes in the brain. Thus, the role of attention in perception/action associations becomes a critical test in determining how important higher cognitive processes are in negotiating these relationships. The H-PAC model I have proposed hypothesizes a negligible role of higher cognitive processes and thus a minimal role for attention. Most other models make similar predictions, with the noteworthy exception of control theory approaches, which may be taken to assume some role of attention and higher-cognitive processes, given their focus on error monitoring.

The way one can test the extent of attentional demands in music perception and action can be through the effects of altered auditory feedback on *executive control*. The concept of executive control in psychology is closely associated with the classic *Working Memory Model* (Baddeley & Hitch, 1974), which is one of the most influential theories in cognitive psychology. Working memory refers to a short-term memory store used to perform mental operations on the information you store. Remembering a series of digits involves only short-term memory, whereas performing mental arithmetic on those digits involves working memory. Baddeley's model of working memory has a hierarchical configuration in which a higher-level *central executive* allocates attention to lower-level components devoted to modality-specific forms of memory. For our purposes a critical component is the *phonological loop*, which was originally proposed as a store for verbal information but may also serve working memory for musical sequences (Salamé & Baddeley, 1989; Schendel & Palmer, 2007; Williamson, Baddeley, & Hitch, 2010; but see Deutsch, 1970). The executive control that this component of memory provides helps us filter out unimportant information to maximize the utility of working memory.

These ideas apply to the concepts of this chapter in the following way. If associations between perception and action rely on higher cognitive functions that require attention, then they will place demands on executive control. As a result, if you are asked to *do something else* at the same time as you play a tune and listen to yourself, your ability to do that other task should suffer. Such a scenario involves a *dual-task* situation. We engage in dual tasks more often than we should. We try to drive and talk on the phone (even a hands-free phone involves a dual task), we watch TV and do our homework, and some of us may even read the newspaper while listening to the news on the radio (yes, I do this, as silly as it sounds). A cruel fact of our evolution is that our brains evolved to focus on one thing a time, and are poorly suited to the dual-task situations that have become a hallmark of modern life. Music performance, even among experts, is not immune to these effects. Opera singers (Çorlu, Maes, Muller, Kochman, & Leman, 2015), cellists (Maes, Wanderley, & Palmer, 2015), and drummers (Fischinger, 2011) exhibit disrupted performance timing when given a memory-demanding second task. But the question for now must focus on music performance, and in particular whether *alterations* of auditory feedback cause us to engage executive control processes as a result. In fact Krampe and colleagues (2005) proposed that sequential planning in general does require executive control, whereas low-level timing may not. If so, we may expect effects related to executive control when the sequential content of feedback is altered, but not timing.

What happens to effects of altered feedback in a dual-task paradigm? If participants rely on executive resources during these tasks, then the adding more demands should enhance the disruptive effect of altered feedback relative to performance with normal auditory feedback. This was found in an experiment reported by Pfordresher and Beasley (2014), which was also discussed in Chapter 5. In this experiment, participants performed sequences from memory while experiencing normal or serially shifted feedback (see Chapter 5 for more details of the feedback alteration). On half the trials participants were given a secondary task that was designed to challenge serial order memory (Milton, Longmore, & Wills, 2008). Before the trial, the experimenter read a sequence of six digits to the participant that needed to be retained in memory during the trial. At the end of the trial, participants would be asked to recall the number for a specific serial position, using the preceding number as the cue (e.g., "What number came after the number 5?"). Although the need to complete this task had no effect while performing with normal auditory feedback—participants could perform the piece and also respond to numbers with high accuracy—error rates during altered feedback conditions increased in dual-task trials. Interestingly, performance on the secondary digit memory task also declined during altered auditory feedback trials. This result suggests that performance with altered feedback, though already generally poor, may worsen when attention is diverted.

What happens when a manipulation is designed to *increase* attention to auditory feedback? As mentioned previously, participants typically try to ignore auditory feedback. It may be the case that altered feedback attracts attention to sound, thus diverting attention from motor performance, leading to disruption. Thus, one is inclined to wonder if disruption may also increase if participants are made to pay attention, for instance by having to count the number of times an additional sound occurs during a performance. We have run an unpublished study of this sort in my lab. Increased attention to sound did not increase disruption; in fact it diminished disruption slightly (though nonsignificantly).

In the domain of speech production, Martin and colleagues (2018) examined whether an individual's ability to adapt to altered feedback is predicted by independent tests of executive control. These authors used alterations of vowel content similar to the study of Houde and Jordan (1998) discussed in Chapter 7. The sort of feedback alteration they used is of interest here because participants can adapt to it by changing the kind of vowel they produce so that auditory feedback still matches the target vowel. But some people are better at adaption to the altered feedback than others. Executive control ability was measured using a set of common tasks, all of which require the participant to ignore potentially distracting irrelevant information. For instance, in the "flanker task," also described in Chapter 5 (Eriksen & Eriksen, 1974), participants see an array of arrow symbols and have to report the direction of the arrow just in the center, while ignoring the arrows that surround it. No measure of executive control predicted participants' adaptation to auditory feedback, in contrast to measures of perceptual acuity (the ability to discriminate pitch and loudness of sounds).

Taken together these results suggest that executive control is not necessary for our processing of auditory feedback, and that demands on executive control may not be responsible for the disruptive effects of altered feedback. Nevertheless, participants may engage in greater executive control as a result of these disruptive effects, and thus may be limited in how well they can accommodate multiple tasks. Although these effects come from admittedly artificial manipulations of feedback, they are related to scenarios in the real world. Consider all those car-chase scenes in movies where the brakes suddenly give out. Prior to that moment, the driver pays no attention to the brakes. However, once perceptual feedback signals that something is amiss, the driver's attention becomes absorbed by the problem of stopping the car. Effects on the car's trajectory, and the driver's behavior, were not directly resulting from how the faulty brakes engaged attention.

INCORPORATING METACOGNITION IN THE H-PAC MODEL

Fig. 8.1 lays out an expanded model of perception/action coordination that incorporates the higher-level features discussed in this chapter, building on the representation shown in Fig. 5.1. Some aspects of this model must be considered hypothetical based on the lack of direct evidence in the existing literatures.

The lower half of this figure illustrates sensorimotor factors in perception and action, which play a critical role in both the experience of agency and error monitoring. The basic sensorimotor loop, shown on

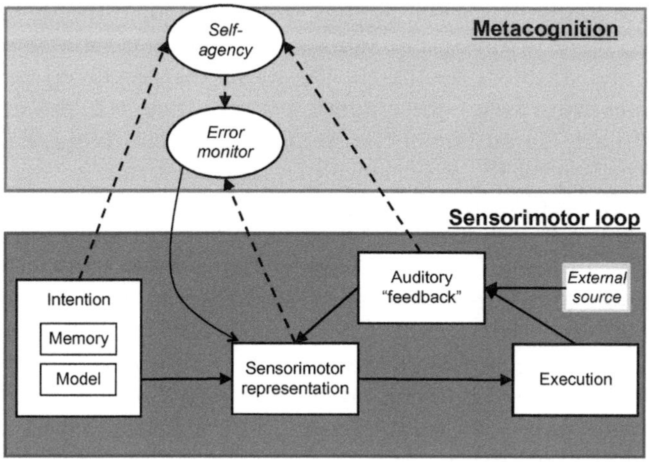

FIGURE 8.1 Relationship between higher cognitive and sensorimotor processes in the H-PAC model.

the lower right side, is consistent with the sensorimotor model described in Chapter 5. As discussed in Chapter 1, intentions are auditory representations of an accurate sequence. The intention is often abstract and hierarchical. During planning, the performer accesses the intention to update plans for actions incrementally to construct a sensorimotor representation that coordinates perception and action. Whereas the intention is thought to represent an ideal performance, the sensorimotor representation may generate errors and also reflects difficulties that can occur during altered auditory feedback or when performing in the presence of other interfering auditory information.

The new aspects here are the higher-level processes, shown as ovals in the upper half of the expanded model. Selfagency (or authorship) is influenced by the match between auditory feedback and the sensorimotor representation used to plan actions. With increasing discrepancy between the auditory representation of ongoing sound, and the shared representation used to coordinate perception and action, agency diminishes.

In this model agency has no direct connection with the sensorimotor representation modeled by H-PAC. This architecture is based on the result that agency is not linearly related to disruption of production (Couchman et al., 2013). The sensorimotor representation responds in a nonlinear way to associations between perception and action, and can "tune out" feedback patterns that are highly discrepant from the planned pattern, as well as sequences that lack salience due to higher-order features like tonality. By contrast, existing evidence on the experience of agency suggests that there is a linear, or at least monotonic, relationship between agency and perception/action similarity. Thus it makes sense to dissociate self-agency from the sensorimotor representation. In Fig. 8.1, I propose that self-agency reflects mismatches between intentions and auditory feedback: the more the perceived pattern you hear deviates from what you intended to produce, the less certain you are that you are the author of the sounds that occur. Note that this prediction accords with the experience of performers in a group, when the sound of a different instrument may mask the sound of your own.

Error monitoring in the model reflects two different processes, as suggested by the data from Pfordresher and Beasley (2014). One source of information comes from the same interaction of feedback with intentions that influence self-agency. This source arises from the fact that performers occasionally mistake altered pitches for errors they themselves generate. At the same time, most registered errors reflect true disruption of production, while also being strongly influenced by veridical disruption from altered feedback.

An important prediction of this model that still needs testing comes from the link between self-agency and error monitoring. According to this model, conditions that decrease the experience of agency should lead to more alterations that are mistaken for errors. At this point a potential problem appears. As mentioned earlier, Pfordresher and Beasley (2014) found no effect on error monitoring, as well as no disruption, from occasional altered pitches that were randomly selected. At the same time, Couchman and colleagues (2013) found that randomly selected pitches lead to minimal experience of agency. This apparent discrepancy can be resolved by noting that the weighting of sensorimotor information in the error monitor is much greater than information from self-agency, which can be thought of as a discrepancy detector.

How does this system operate in real time during a performance? As can be seen, production does not depend on any higher-order processes. However, error monitoring may be used to adjust the sensorimotor representation. Though not explicitly represented in Fig. 8.1, I propose that these adjustments operate based on a threshold function. For low-levels of error, the sensorimotor loop continues unabated, given standard practices in music performance described earlier. However, if more extreme disruption happens (e.g., a piano key starts consistently producing the wrong pitch), the error monitor may be used to make adjustments.

CONCLUDING REMARKS

Of all the topics discussed in this book, the way in which higher-order cognitive processes influence associations between perception and action is the least well explored. Yet, the relatively small number of existing studies allow us to draw some tentative conclusions. Although the effects of auditory feedback on performance seem to operate autonomously and do not require any higher cognitive processing, clearly these higher processes influence and are influenced by perception and action. Although altered feedback does not need to sound like an error to disrupt production, neural and cognitive error monitoring processes do respond to these alterations. Even more intriguing is the apparent nonlinear relationship between our sense of authorship for auditory feedback and the disruptive effect that can occur when the pitch content of feedback is altered. Finally, although altered feedback can be distracting (as we see in Gonzo's weekly gauntlet), attempts to ignore or attend to auditory feedback seem to have little effect on its potential to disrupt.

Certain paradigms in this chapter may make the reader wonder what happens in situations where the auditory environment includes not only one's own auditory feedback, but also the sounds of other performers. If, as I have claimed, sensorimotor processes relating perception and action do not depend on higher cognition, one would conclude that we treat the sound of other musicians the same way as the sound of ourselves. This seems counterintuitive. In the next chapter I discuss research concerning whether this counterintuitive prediction holds.

CHAPTER

9

Coordinating With Others

To this point, we have considered perception/action associations for solo performances only. Several reasons justify this limitation. Solo performance constitutes a significant portion of music, with many salient contributions (most of Chopin's compositions, Jimi Hendrix playing the National Anthem, etc.). Also, performers who attain high levels of expertise typically spend many hours practicing on their own, fine-tuning sensorimotor associations in the process. Finally, solo performances offer the scientist much better control than performances in duos or groups, where an array of social variables, as well as compounding error variance across individuals, can influence results. In short, solo performances make a good starting point for understanding the role of sound in music performance. But let's face it—most music making is done by duos (pairs of musicians) and groups (three or more). Thus we must expand our perspective.

Something magical happens when you perform music with other people and the group develops an implicit sense for each other. The combined voices of Simon and Garfunkel blended together have an emotional force that neither singer on his own (in my opinion, at least) quite matches. Together, the duo or group becomes more than the individuals and a special kind of unity forms. This chapter concerns how individuals in such a collective activity respond to each other.

For the purpose of this book, a major question concerns whether the role of sound produced by another individual has a similar influence to the role of your own auditory feedback. Results discussed in earlier chapters show that alterations of auditory feedback can be disruptive even if they are not interpretable as the results of your own intentions. If so, then one might expect the sound pattern being produced by the violist to influence the cello player's performance in the same way as would alterations to the violinist's own auditory feedback. Nevertheless there are important differences to consider when translating from

Sound and Action in Music Performance
DOI: https://doi.org/10.1016/B978-0-12-809196-8.00009-7

research on solo performance to research on what is called *joint performance*, in which multiple musicians produce music cooperatively.

One major difference between solo and joint performance is that in joint performance there are multiple sources of information available, including one's auditory feedback as well as sounds produced by other individuals. Thus, in joint performance there is the issue of how one processes auditory input as a function of the source of that input. As we discussed in Chapter 4, the brain may suppress access to one's own auditory feedback (Pfordresher, 2014), thus dampening its potential to obscure feedback from the other individual.

Another difference is that in joint performance the music one intends to perform comes from the integration of all parts together. Thus, even though the sound patterns produced by other musicians may be difficult to coordinate with your own plan, all parts matter and a good musician learns to coordinate while attending to all parts (cf. Keller, 2001). By contrast, in studies using altered auditory feedback the sounds one hears offer no utility whatsoever. You always know the weak singers in a choir because they are more likely to block the sound of their neighbor by pushing their earlobe into their own ear canal. The stronger singers thrive on hearing their own part alongside others.

Third, in joint performance all parts are typically not created equal. Most piano duets designate the *primo* performer as conveying the melody while the *secondo* performs an accompaniment. In jazz ensembles, performers take turns at the helm, soloing one after the other in a cycle. Even in barbershop quartets, where singers often generate block chords, one of these parts draws our attention. Thus, the musical structure itself contains a kind of hierarchy. Beyond this, performers are humans and may be influenced by their relationship to other performers as humans. During the recording of their famous last album, *Synchronicity*, The Police (a rock band from the early 1980s) reportedly had so many interpersonal problems that recordings of the band members were conducted in isolation (Buskin, 2004)!

Finally, whereas it has been possible so far to pay little attention to visual feedback, we cannot do this any longer. Visual feedback may play a minor role in most solo performance. Many instrumentalists cannot clearly see what they are doing (e.g., the saxophone), and even when it is possible to do this a skilled musician will not watch their own actions (e.g., a concert pianist). In joint performances, however, visual information matters critically, as any youth orchestra conductor can attest.

Although joint performance requires musicians to converge on many dimensions, including the intensity of sound and the intonation of pitch, my main concern here will be on synchronization. To date synchronization has been the primary focus of research, and for performing musicians this dimension arguably poses the biggest challenge. Orchestras use conductors in large part to remain synchronized. Drummers serve the same purpose in rock bands, bassists in jazz, and in smaller ensembles the leader will nod his or her head to convey the beat. The main issue for this chapter will be how musicians use auditory information from other performers to synchronize, although I will touch on these other important points along the way.

THINKING ABOUT MUSIC TOGETHER: JOINT REPRESENTATIONS

A major theoretical issue in the study of joint performance is whether performers rely on a *joint representation*, defined as "... [a] shared performance goal, that is, a unified conception of the ideal integrated ensemble sound" (Keller, Novembre, & Loehr, 2016, p. 283). Critically, a joint representation presumes knowledge about the musical structure of all musical parts involved in a performance, as well as any expressive nuances that may also play a part in the execution. Of course, the degree of clarity for different parts may vary. Two orchestra members performing in unison know exactly what their partner wishes to play. Two pianists playing a duet most likely read from the same notation. By contrast, the trumpeter in an orchestra may not have such a clear understanding of the cello part, and jazz instrumentalists may have only a very rough idea of what their improvising bandmates may do.

A laboratory in Hungary, headed by Natalie Sebanz and Gunther Knöblich, has taken a lead role in advocating for the role of joint actions (Sebanz, Bekkering, & Knoblich, 2006; Sebanz & Knoblich, 2009). A former postdoc of this lab, Janeen Loehr, applied this concept to musical tasks both in Hungary and her present lab at the University of Saskatchewan. Together this team published a convincing demonstration that pianists do form joint representations of their part and their companion's part, and that this joint representation integrates perception and action (Loehr, Kourtis, Vesper, Sebanz, & Knoblich, 2013).

Before exploring the results of this study further, let's consider what a joint representation entails. Fig. 9.1 provides a cartoon illustration of this concept. First, a joint representation involves knowledge of your part as well as those of other performers. These constitute *solo representations* in Fig. 9.1. As noted earlier, this knowledge may be unevenly distributed in practice. Laboratory studies of joint representations usually simplify the matter by having participants either fully knowledgeable or not knowledgeable at all about parts other than

FIGURE 9.1 Cartoon of solo versus joint representations in a piano duet performance.

their own. Second, although a joint representation needs to include information about all parts, these parts must be differentiated such that the performer does not mistakenly produce another performer's part. In this respect, joint representations may involve differential weighting and/or a hierarchical organization in which parts are unified at a higher level of organization and differentiated at a lower level. Finally, the joint representation must integrate perception and action. When performing together, a joint representation causes you to predict what your coperformers will do next. This need not be limited to a prediction of the sounds they will create; there is evidence that the motor system responds in anticipation of what coperformers will do next in generic experimental psychology tasks, like responding to images on a computer screen (Dolk et al., 2014). Of course, motor skill learning may limit this tendency in some musical situations—my own ability to anticipate motor behavior in a saxophonist is rather limited—but among pianists playing a duet there is considerable possibility for motor interactions. And that is what Loehr and colleagues addressed.

Loehr and colleagues (2013) used manipulations of auditory feedback to explore how intertwined the representations of self and other are during performance. This study capitalized on a unique feature of musical duets: the fact that in duets there are separable individual and joint goals, and both goals are highly salient to each performer. Each individual part in this study was a sequence of two-note chords. Both parts combined to form a sequence of four-note chords creating a rich harmonic texture. Manipulations of auditory feedback that altered the harmonic structure would thus disrupt both the individual and the joint representation of planning, whereas an alteration that did not change the harmony would only disrupt the individual representation.

The authors measured electroencephalography (EEG) and focused on two feedback components, a negative-going event-related potential that follows error-related feedback at a latency of about 200 ms (also measured by Maidhof et al., 2010; see Chapter 4), as well as a positive-going response at a latency of around 300 ms, a higher level response that reflects evaluation of stimuli relative to one's working model of the environment (Donchin & Coles, 1988). Whereas all alterations led to the earlier error feedback-related response, the later "P300" response varied across conditions. In general, the P300 was larger for alterations to one's own feedback rather than the coperformer, which suggests prioritization of one's own part in a duet. More important for the joint representation view, the P300 was also larger when alterations influenced the joint goal, that is, alterations that disrupted the harmony, as opposed to disruptions that only influenced one performer's goal. This result is important because it suggests that the significance of feedback alterations in a duet are based on a representation that integrates the action plans for both performers.

Another paper that addressed the neural action/sound associations in the joint representation came from Peter Keller's group at the Max Planck Institute in Leipzig (Novembre, Ticini, Schütz-Bosbach, & Keller, 2012). In this study, the authors had pianists practice two-hand pieces, and then perform just the right-hand part in a test phase. The practice phase ensured that pianists had a joint action plan based on auditory and motor knowledge of both parts, even though only one part was planned for execution at test. During performances the authors used *transcranial magnetic stimulation* (TMS), which was described in Chapter 2, to probe motor reactivity of the unperformed part. The authors found that TMS elicited contractions in the participants' left forearm during performances, as if performers were on the verge of executing a movement. Importantly, no such activity was found if pianists performed a melody with the right hand while hearing an accompanying part they had never played or heard before. These results support the joint action hypothesis in that performers appear to have a motor plan at the ready for a musical part they have encoded as part of the joint action plan and yet do not intend to play. At the same time, there is an important qualification in their results. The effects I just described were also found when pianists did not hear any auditory input from the accompanying part. Thus, the fact that they might imagine the presence of an accompanying left hand was enough. This is highly important for our broader concern with the role of sound in music performance, as it suggests a potentially limited role for auditory information in a joint action plan. A subsequent study further supported this view, by showing weaker coupling of timing between dueting pianists when each performer knew the other performer's part, in contrast to a condition in which each performer was unfamiliar with the other part (Novembre, Sammler, & Keller, 2016).

These papers suggest that performers do form joint perception/action representations in the brain. But what role do these representations play in performance? The most likely function of joint representations would be to facilitate synchronization across performers. However, as noted above, effects on synchronization of tone onsets does not support this position. However, it is possible that facilitatory effects of joint representations facilitate synchronization at broader timescales. When we get to know a piece of music, our understanding of its structure deepens. The piece ceases to be conceptualized as a series of isolated notes. Instead, the notes form patterns that communicate melodic and harmonic motion, with phrase boundaries separating different segments. When we gain an understanding of our coperformer's part, we are then able to coordinate actions with that performer at these broader timescales more effectively. The ability to synchronize individual tones, which is what researchers have traditionally measured, may not capture the effect of such higher-level knowledge.

Another study from the Max Planck Institute addressed this hypothesis (Ragert, Schroeder, & Keller, 2013). In addition to the use of standard synchronization measures for tone onsets, this group also used *motion capture* to evaluate synchronization at broader timescales. Motion capture measures body movements by recording changes in position of small "markers" that are placed on the body. Depending on the setup these markers may emit or reflect light. When markers are placed appropriately, one can measure movement of the head, torso, arms, etc. in real time and get a sense for the position, velocity, and acceleration patterns of these movements. Motion capture, which music performance researchers started using in the early part of the 21st century, allows researchers to evaluate performance timing in a much more comprehensive way than simply looking at the timing of tone onsets. In addition, many body movements respond to broad timescales of movement, like the beat and metrical accents; this has been found in research on free-form dance (Burger et al., in press). Ragert and colleagues (2013) found that knowledge of the coperformer's part did not facilitate synchronization of individual notes. However, this kind of joint knowledge did facilitate higher-level synchronization based on vision that tracked movement at large timescales.

MAINTAINING COHESION THROUGH SENSORIMOTOR ENTRAINMENT

One limitation of the joint representation view is that it does not provide a clear mechanism for *how* performers synchronize with each other, once a mutual concept of performance has been constructed. When your coperformer hits a key at a time when you didn't quite expect it, how do you adapt to get back on course? This sort of account has been dealt with by another class of models that focus on lower-level features of performance. However, there is one particular approach that some have put forth as an alternative to the joint representation view, which I here refer to as the *entrainment* model.

The entrainment model flows from the *Dynamical Systems* approach discussed in Chapter 3. Entrainment refers to as the process of synchronizing two ongoing rhythms based on sensory input. Typically these rhythms are modeled as ongoing cycles, or oscillations, and so these models are sometimes called "oscillator models." Fig. 9.2 illustrates various dimensions on which entrainment can occur in a piano duet performance. The most well-studied form of entrainment in music is the synchronization of joint actions based on auditory feedback. However, in most performances one can also use visual information, as discussed in the previous section.

One potentially attractive feature of entrainment models is that they do not require any knowledge about the coperformer's part—they

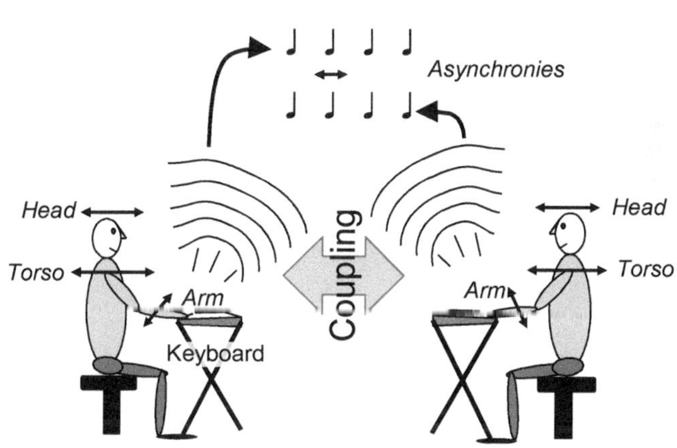

FIGURE 9.2 Cartoon illustrating patterns of coupling in entrainment. Salient rhythmic body movements are highlighted using double-sided arrows. Sound is represented as an acoustic stimulus from each keyboard using wave patterns, and as a perceptual representation via notation above the performers. Coupling may occur across any form of movement, or feature of the sound pattern, as represented by the double-sided arrow in the center.

assume no joint representation. Thus, from the principle of parsimony entrainment models have an advantage over joint representation. Note that the fact that these models make no assumptions about representations does not negate the possibility that performers may develop and use them. Parsimony is an advantage insofar as it does not require the inclusion of some component to account for a phenomenon. This is a particular advantage when one considers the notion of a representation, which is abstract and thus can be difficult to falsify through scientific research.

What distinguishes the entrainment model is its proposal that synchronization involves the unification of two ongoing rhythms.[1] This leads to the somewhat radical proposal that performers have their own intrinsic rhythms, referred to as *endogenous* rhythms, meaning within the organism, as opposed to an *exogenous* rhythm from a stimulus. Entrainment involves adapting an endogenous rhythm to synchronize with the exogenous rhythm. An endogenous rhythm may have a physical basis—the natural frequency with which we swing our arms is slower than the frequency at which we may wiggle our index finger—but may be more abstract and internal. A large amount of psychological research, for instance, suggests that individuals exhibit a preferred spontaneous rate at which they produce rhythmic movements, and that this may vary across individuals or within an individual as they age (getting slower, as one might expect, McAuley et al., 2006). Endogenous rhythms are not only manifested in musical behaviors, but may be seen in rates of walking and talking too.

An entrainment model predicts that performers synchronize best with each other when their endogenous rhythms match. Researchers typically measure endogenous rhythms through a performer's spontaneous tempo. This is as simple as it sounds. The performer enters the lab, and performs a piece of music at whatever tempo they choose, the only constraints being that it is a "comfortable" tempo, not super fast or languidly slow.

Convincing support for this hypothesis was reported in a study carried out by Anna Zamm at the Sequence Production Lab at McGill University, led by Caroline Palmer (Zamm, Wellman, & Palmer, 2016). Participants in this study visited the lab twice. During their first visit, the researchers recorded the spontaneous tempo at which each pianist performed. Afterwards each pianist was paired with another pianist, based on one of two grouping strategies. In one group, pianists were paired based on the best possible match in spontaneous tempo. For this group, the endogenous rhythms of each performer matched the

[1]Not all models of synchronization involve entrainment in the dynamical systems sense. Alternate models may be based on linear error correction (e.g., Vorberg & Wing, 1996).

other quite well based on previous measures of spontaneous tempo, and as a result synchronizing these rhythms in a duet ought to be straightforward. In the other group, each pianist was paired with another pianist who had a highly dissimilar spontaneous rate: all pianists in the fastest half of the distribution were paired with a pianist in the slowest half. The endogenous rhythms across individuals in a pair were thus not well matched and so synchronizing in a duet ought to be more difficult. This is what the authors found. Pairs that were formed based on a close match of spontaneous tempo synchronized together more effectively than pairs that were formed based on a poor match. This study stands out because it used random assignment to pair participants based on spontaneous tempo. Previous studies had shown a similar effect by correlating tempo matches in solo performances with synchronization in duet performances within a single session (Loehr & Palmer, 2011; Zamm, Pfordresher, & Palmer, 2015).

Because entrainment models emphasize the role of sensorimotor coordination, as opposed to cognitive representations, entrainment models predict that the presence of auditory feedback should have a strong role in joint coordination. Several studies from Caroline Palmer's lab support this view. Duet synchronization suffers significantly when one performer cannot hear the other performer, even when they can see each other (Goebl & Palmer, 2009). Likewise, when you can *only* hear the other performer, synchronization suffers too (Zamm et al., 2015). Not surprisingly, synchronization suffers the most if no auditory feedback is available from either performer in a duet (Demos, Carter, Wanderley, & Palmer, 2017).

When auditory feedback is present, entrainment models predict a strong influence of the temporal relationship between each performer's part. A coperformer whose part is not perfectly synchronized with your own part will pose a challenge for the process of entrainment. Goebl and Palmer (2009) measured synchronization for pairs in which one part may have a rate that is double the other part. Although this kind of temporal relationship is not highly complex, it did prove more challenging to performers than unison performances, particularly when auditory feedback was removed from one of the performers. Another study that addressed how performers synchronize to a metronome—a task that is roughly analogous to duet performance—found that pianists tend to anticipate the beat more if metronome clicks subdivided the time intervals between produced notes (Loehr & Palmer, 2009). An important topic for future research has to do with temporal relationships that are more complex than the 2:1 ratios used so far. According to entrainment models, synchronization based on such simple harmonic ratios (i.e., ratios that can be reduced to an integer) should come much more easily than other forms of synchronization, such as the complex ratios often used in Chopin piano pieces.

INTEGRATION: ENTRAINMENT OF JOINT REPRESENTATIONS

When confronted with two models that explain a phenomenon, one naturally wishes to pit them against each other to see which model succeeds over the other. Such inclinations arise naturally from the important logic of falsification in science (Platt, 1964). However, in many cases different models address a phenomenon at different levels of explanation. That may be the case with joint representation and entrainment models of joint performance. Whereas the joint representation view establishes each performer's expectation and determines how performers relate to each other at broad structural levels (phases, meter, etc.), the entrainment model provides a mechanism for how performers maintain synchronization using online sensory input.

One of the aforementioned papers from Peter Keller's group used different manipulations to tease apart the influences of joint representations versus sensorimotor entrainment (Novembre et al., 2016). The authors manipulated accessibility of performers to a joint representation by varying the familiarity of performers with the coperformer's part. Lower-level sensorimotor entrainment was then manipulated by varying the compatibility of timing across performers. In some cases, the performers had similar instructions for timing, whereas in other cases performers had conflicting instructions (one sped up whereas the other slowed down). Although instructions may constitute part of a representation, the primary influence of instructions was on timing and thus the auditory pattern to which each performer synchronized. Accuracy of synchronization was strongly influenced by low-level auditory information, consistent with entrainment models. However, neural alpha rhythms, measured via EEG, responded strongly to the presence of a joint representation. Alpha rhythms are prominent during awake but inattentive conscious states, but may have other roles to play as well. When performers knew each other's parts (i.e., when they had formed a joint representation), alpha rhythms were suppressed during performances in which both individuals had similar timing instructions, and increased when timing was incongruous. However, no effect on alpha rhythms emerged if performers did not know each other's parts. The authors concluded that the suppression of alpha rhythms might function as a marker for the integration of sensory information from yourself and others around you.

Another possible point of convergence has to do with the role of pitch patterning in joint synchronization. Studies of piano duets have shown that synchronization accuracy is influenced by pitch sequencing in each part. For instance, the aforementioned study from Zamm and colleagues (2015) compared unison performances with performances of the same pieces in a round. Rounds are particularly interesting here because they constitute a real-world analogue to serially shifted feedback, used in research discussed earlier (e.g., Pfordresher, 2003a, 2005, 2008). In a round, your coperformer's part exactly replicates the planned sequence of your own part, but at a lag of several events (usually one or two measures). Thus there is potential for confusion based on pitch sequencing, as well as a shift of higher-order time structure based on phrasing. Thus, not surprisingly, synchronization during performances of rounds is less accurate than synchronization of unison performances. In another study cited earlier, pianists performed duets for which the parts had a simple or relatively complex harmonic relationship to each other (Loehr & Palmer, 2011). Although synchronization across performers was not affected by this kind of complexity, fine-grained analyses of timing within each performer suggested that they timed their actions in ways that highlighted phrase boundaries, possibly to help communicate the complex structure more effectively to their coperformer.

A third way that representational and entrainment views have converged in the literature has to do with the effect of hierarchical relationships in joint performance. Most forms of joint performance give different roles to different players, and typically one part takes a dominant role over others. In classical ensembles, the melody usually takes the lead. One can see this in piano duets (*primo* vs *secondo* parts), as well as other small ensembles like string quartets and *a cappella* singing groups. The leading role stems from the fact that listeners attend to the melody, but leaders take on more responsibility than simply bearing the brunt of the attentional responsibility. Leaders keep time for the group, and also define the reference pitch for tuning, when applicable.

Clearly, the allocation of roles in joint performance ties into a joint representation for the shared goals of all the performers. However, the resulting hierarchy ties into a long-standing modeling scheme from entrainment. Coupled dynamical systems typically distinguish *driving* oscillations from *driven* oscillations, and most of these systems have no high-level cognitive schemas determining their behavior (e.g., changes in heart rate driven by a pacemaker). As the terms suggest, a driving oscillator governs the behavior of a driven oscillation. When applying this to musical contexts, the beat of your favorite song may function as a driving oscillator as you dance. You adapt your own internal rhythm to the beat of this song; thus, your internal rhythm (your favored, most comfortable rate of movement) is driven by (coupled to) the beat.

In the same way, the *secondo* part in a piano duet may be driven by the *primo* part.

As it turns out, piano duet performance may be more egalitarian than one would think. Goebl and Palmer (2009) measured the degree to which each performer's pattern of timing reacted to the timing of the other performer. Specifically, if the *secondo* performer reacts to the *primo*, then the *secondo* will produce a relatively long interval between key-presses just after the *primo* slows down, and vice versa. As a result the pattern of interonset intervals produced by the *primo* for intervals 1–8 should correlate with produced intervals 2–9 in the *secondo*. Goebl and Palmer did find this pattern of timing. However, they also found exactly the reverse: the *primo* player also reacted to timing in the *secondo* part. Piano duets thus exhibit mutual adaptation, not just adaptation of the follower to the leader.

A major way that social roles may be realized is not via the synchronization of sound events but instead in the synchronization of body movements. Analyses of movement via motion capture suggest that the head and torso movements of *primo* performers in a piano duet anticipate those of the *secondo* (Bishop & Goebl, 2018; Keller & Appel, 2010; Goebl & Palmer, 2009). In string quartets, bow movements from the first violin play may play a similar role (Timmers, Endo, Bradbury, & Wing, 2014). Recent analyses of movements in string quartets suggest that the leader's movements significantly predict subsequent movements of other performers (Chang, Livingstone, Bosnyak, & Trainor, 2017).[2] These anticipatory movements may facilitate synchronization of both performers (Keller & Appel, 2010).

NEURAL MECHANISMS

Earlier I discussed a study that used TMS to test the hypothesis that coperformers rely on a joint representation (Hadley, Novembre, Keller, & Pickering, 2015). In that study, TMS was applied over the primary motor cortex. It is not surprising that the primary motor cortex would be important for joint performance, given that any kind of fine-grained motor control must involve the primary motor cortex. What was significant about that study was that TMS elicited muscle contractions for a musical part that the coperformer was not planning to produce. This result suggests that joint representations can have an influence on very low-level features of motor planning.

[2]An important consideration is the type of analysis used to determine how temporal features of one part may predict those in another part. For further discussion see Dean and Dunsmuir (2016).

Two studies from Peter Keller's lab have addressed neural mechanisms that help coperformers maintain synchronization (Fairhurst, Janata, & Keller, 2013, 2014). For purposes of experimental control, these studies had participants tap, rather than play piano, along with a "virtual" partner. The virtual partner was in fact a computer program designed to produce metronome sounds that could adapt to the participants' tapping or could be set to be unresponsive, or even counteract the participant's timing (e.g., if the participant slows down, the program may speed up). The purpose of these manipulations was to determine those brain areas that respond to differences in the entrainment experience. Several areas responded to differences in the pattern of adaptability exhibited by the virtual partner. The most critical ones include motor planning areas, such as the premotor cortex and the inferior frontal gyrus.

When people first encounter the entrainment model, the concept of an internal rhythm can seem flighty. One is inclined to question whether there are "oscillators in the brain." Well, it seems that there are! We have known about neural oscillations (a.k.a. "brain waves") for some time, but now we know that oscillations can adapt to rhythms of music (for a review, see Haegens & Zion Golumbic, 2018). Even more astounding, when musicians play duets their neural oscillations may synchronize with each other. Lindenberger and colleagues had guitarists play duets while brain activity was measured via EEG (Lindenberger, Li, Gruber, & Müller, 2009). Neural oscillations across musicians synchronized in phase with each other, even before the performers started to play. Other more recent research on piano duets verified that neural oscillations with a frequency that closely matches the tempo of the performance are more prominent that oscillations at other frequencies (Zamm et al., 2018).

The causal importance of this neural synchronization was shown in a study of joint tapping (a simpler task than piano duets, Novembre, Knoblich, Dunne, & Keller, 2017). Novembre and colleagues had participants tap in synchrony with each other, and during some trials the participants would experience periodic electrical stimulation of the area of motor cortex responsible for controlling the movements. The technique the authors used is called *transcranial alternating current stimulation*; the stimulation involved is low voltage and painless (analogous to TMS, mentioned earlier). The authors varied the rate of pulses, and found facilitation of joint synchrony for oscillations within the range of "beta waves" in the electroencephalograph (i.e., around 20 Hz). This complements the earlier study showing that alpha band activity is suppressed (Novembre et al., 2016). Beta band activity is also enhanced during performances with alterations to auditory feedback content that disrupt the production of serial order (Herrojo Ruiz et al., 2017).

A different kind of joint performance was addressed in a neuroimaging study carried out at Johns Hopkins, led by Charles Limb—an otolaryngologist with a keen interest in creative process of music performance (Donnay, Rankin, Lopez-Gonzalez, Jiradejvong, & Limb, 2014). Whereas in classical music it is common for musicians to play prerehearsed parts simultaneously, in jazz there is a tradition that is closer to conversational speech, called "trading fours." In this format, two musicians alternate short improvised segments that last for four measures each. Like speech, these musical segments are spontaneous and the parts alternate rather than overlap. A critical hypothesis posed by this research was whether this speech-like alternation would yield brain activity in speech planning areas—which is what the authors found. In particular, as in other studies of joint performance, the inferior frontal gyrus was active in both left and right hemispheres. The presence of left hemisphere activity in this area is significant given that it is part of Broca's area, which for some time was considered speech-specific, but now must be considered to have a more general role given many studies, including Donnay and colleagues, that show activity in this area during musical tasks (e.g., Maess et al., 2001).

CONCLUDING REMARKS

In this chapter I summarized results from a rapidly growing area of research in music performance: the way in which performers collaborate when they perform together. This area is compelling in multiple respects, including its ecological validity for the study of music and the fact that it signals a new synthesis between cognitive and social traditions within psychology.

A critical question in this chapter is whether the role of sound in music performance differs for joint performances as opposed to solo performance. In a trivial sense the answer has to be "yes." As pointed out at the start of this chapter, joint performance involves the coordination of sounds resulting from the actions of multiple individuals. This poses a greater perceptual demand than individual performance. In addition, performers assign different priorities to sound patterns as a function of their source. One's own feedback plays a different role than information from others, and successful joint performance involves coordinating movements to optimize coordination across all the sound sources.

A less obvious corollary of the question articulated above is whether the sound patterns attributed to other performers yield a different effect on performance than sounds attributed to you. Although intuitively it may seem as though the answer here must be "yes" as well, evidence to date is not so clear. Recall that in Chapter 8 I discussed how the experience of agency and authorship (i.e., the sense that you are the source of your auditory feedback) did not strongly influence the effect of altered auditory feedback. In duet performance research, there is evidence that being able to hear a coperformer has a greater influence than being able to hear yourself, as one might expect. At the same time the mechanisms that determine how we coordinate with another person based on auditory feedback may not differ from the way in which we respond to our own auditory feedback. Both Chapter 3 and the present chapter invoked the concept of entraining internal rhythms to external auditory events as an account for interactions between perception and action during music performance. In that respect, joint performance may simply involve the use of lower-level sensorimotor functions complemented by higher-order joint representations that determine how we prioritize different sound sources.

References

Acevedo, S., Temperley, D., & Pfordresher, P. Q. (2014). Effects of metrical encoding on melody recognition. *Music Perception: An Interdisciplinary Journal, 31*, 372–386.

Albouy, P., Weiss, A., Baillet, S., & Zatorre, R. J. (2017). Selective entrainment of theta oscillations in the dorsal stream causally enhances auditory working memory performance. *Neuron, 94*, 193–206. e195.

Ammirante, P., & Thompson, W. F. (2012). Continuation tapping to triggered melodies: Motor resonance effects of melodic motion. *Experimental Brain Research, 216*, 51–60.

Aschersleben, G., & Prinz, W. (1997). Delayed auditory feedback in synchronization. *Journal of Motor Behavior, 29*, 35–46.

Aslin, R. N., & Newport, E. L. (2012). Statistical learning: From acquiring specific items to forming general rules. *Current Directions in Psychological Science, 21*, 170–176.

Ayotte, J., Peretz, I., & Hyde, K. (2002). Congenital amusia: A group study of adults afflicted with a music-specific disorder. *Brain, 125*, 238–251.

Baddeley, A. D., & Hitch, G. J. (1974). Working memory. In G. A. Bower (Ed.), *The psychology of learning and motivation* (pp. 47–89). Academic Press.

Badets, A., Koch, I., & Philipp, A. M. (2016). A review of ideomotor approaches to perception, cognition, action, and language: Advancing a cultural recycling hypothesis. *Psychological Research, 80*, 1–15.

Balasubramaniam, R., Wing, A. M., & Daffertshofer, A. (2004). Keeping with the beat: Movement trajectories contribute to movement timing. *Experimental Brain Research, 159*, 129–134.

Bangert, M., & Altenmüller, E. O. (2003). Mapping perception to action in piano practice: A longitudinal DC-EEG study. *BMC Neuroscience, 4*, 26.

Bangert, M., Peschel, T., Schlaug, G., Rotte, M., Drescher, D., Hinrichs, H., et al. (2005). Shared networks for auditory and motor processing in professional pianists: Evidence from fMRI conjunction. *NeuroImage, 30*, 917–926.

Baumann, S., Koeneke, S., Schmidt, C. F., Meyer, M., Lutz, K., & Jäncke, L. (2007). A network for audio-motor coordination in skilled pianists and non-musicians. *Brain Research, 1161*, 65–78.

Beck, S. L., Rieser, J. J., & Erdemir, A. (2017). Singing without hearing: A comparative study of children and adults singing a familiar tune. *Psychomusicology: Music, Mind, and Brain, 27*, 122–131.

Belyk, M., Pfordresher, P. Q., Liotti, M., & Brown, S. (2016). The neural basis of vocal pitch imitation in humans. *Journal of Cognitive Neuroscience, 28*, 621–635.

Bendixen, A., SanMiguel, I., & Schroger, E. (2012). Early electrophysiological indicators for predictive processing in audition: A review. *International Journal of Psychophysiology, 83*, 120–131.

Beranek, L. L. (2004). *Concert halls and opera houses: Music, acoustics and architecture*. New York: Springer.

Berkowska, M., & Dalla Bella, S. (2009). Acquired and congenital disorders of sung performance: A review. *Advances in Cognitive Psychology, 5*, 69–83.

Berkowska, M., & Dalla Bella, S. (2013). Uncovering phenotypes of poor-pitch singing: The Sung Performance Battery (SPB). *Frontiers in Psychology, 4*, 714.

Bidelman, G. M., & Alain, C. (2015). Musical training orchestrates coordinated neuroplasticity in auditory brainstem and cortex to counteract age-related declines in categorical vowel perception. *The Journal of Neuroscience, 35*, 1240–1249.

Bidelman, G. M., Gandour, J. T., & Krishnan, A. (2011). Musicians and tone-language speakers share enhanced brainstem encoding but not perceptual benefits for musical pitch. *Brain and Cognition, 77*, 1–10.

Bidelman, G. M., Hutka, S., & Moreno, S. (2013). Tone language speakers and musicians share enhanced perceptual and cognitive abilities for musical pitch: Evidence for bidirectionality between the domains of language and music. *PLoS One, 8*, e60676.

Bishop, L., Bailes, F., & Dean, R. T. (2013). Musical imagery and the planning of dynamics and articulation during performance. *Music Perception, 31*, 97–117.

Bishop, L., Bailes, F., & Dean, R. T. (2014). Performing musical dynamics: How crucial are musical imagery and auditory feedback for expert and novice musicians? *Music Perception, 32*, 51–66.

Bishop, L., & Goebl, W. (2018). Communication for coordination: Gesture kinematics and conventionality affect synchronization success in piano duos. *Psychological Research, 82*, 1177–1194.

Black, J. W. (1951). The effect of delayed side-tone upon vocal rate and intensity. *Journal of Speech and Hearing Disorders, 16*, 56–60.

Black, J. W. (1955). The persistence of effects of delayed side-tone. *Journal of Speech & Hearing Disorders, 20*, 65–68.

Boersma, P., & Weenink, D. (2013). *Praat: Doing phonetics by computer (Computer program, Version 5.4.09)*. Retrieved from: <http://www.praat.org/> 25.08.14.

Boltz, M. G., & Jones, M. R. (1986). Does rule recursion make melodies easier to reproduce? If not, what does? *Cognitive Psychology, 18*, 389–431.

Botvinick, M. M., & Cohen, J. (1998). Rubber hands 'feel' touch that eyes see. *Nature, 391*, 756.

Bregman, A. S. (1990). *Auditory scene analysis: The perceptual organization of sound*. Cambridge, MA, US: The MIT Press.

Brown, G. S., & Campbell, D. P. (1948). *Principles of servomechanisms: Dynamics and synthesis of closed-loop control systems*. New York, NY: Wiley.

Brown, R. M., & Palmer, C. (2012). Auditory-motor learning influences auditory memory for music. *Memory & Cognition, 40*, 567–578.

Brown, R. M., & Palmer, C. (2013). Auditory and motor imagery modulate learning in music performance. *Frontiers in Human Neuroscience, 7*, 320.

Brown, S. (2007). Contagious heterophony: A new theory about the origins of music. *Musicae Scientiae, 11*, 3–26.

Brown, S., Ingham, R. J., Ingham, J. C., Laird, A. R., & Fox, P. T. (2005). Stuttered and fluent speech production: An ALE meta-analysis of functional neuroimaging studies. *Human Brain Mapping, 25*, 105–117.

Brown, S., Ngan, E., & Liotti, M. (2008). A larynx area in the human motor cortex. *Cerebral Cortex, 18*, 837–845.

Brown, S., Pfordresher, P. Q., & Chow, I. (2017). A musical model of speech rhythm. *Psychomusicology: Music, Mind, and Brain, 27*, 95.

Burger, B., London, J., Thompson, M. R., & Toiviainen, P. (2018). Synchronization to metrical levels in music depends on low-frequency spectral components and tempo. *Psychological Research, 82*, 1195–1211.

Burger, B., Thompson, M., Luck, G., Saarikallio, S., & Toiviainen, P. (2013). Influences of rhythm- and timbre-related musical features on characteristics of music-induced movement. *Frontiers in Psychology, 4*.

Burnett, T. A., Kiran, S., & Larson, C. R. (1997). Voice F0 responses to feedback pitch are not affected by speed of stimulus onset. *The Journal of the Acoustical Society of America, 102*, 3206–3206.

Buskin, R. (2004). Classic tracks: The police's 'every breath you take', March *Sound on Sound*, 136–140.

Butler, R. A., & Galloway, F. T. (1957). Factoral analysis of the delayed speech feedback phenomenon. *The Journal of the Acoustical Society of America, 29*, 632–635.

Carter, C. S., Braver, T. S., Barch, D. M., Botvinick, M. M., Noll, D. C., & Cohen, J. D. (1998). Anterior cingulate cortex, error detection, and the online monitoring of performance. *Science, 280*, 747–749.

Chaffin, R., Demos, A. P., & Logan, T. R. (2016). Performing from memory. In S. Hallam, I. Cross, & M. Thaut (Eds.), *The Oxford handbook of music psychology* (2nd ed., pp. 559–572). Oxford: Oxford University Press.

Chang, A., Livingstone, S. R., Bosnyak, D. J., & Trainor, L. J. (2017). Body sway reflects leadership in joint music performance. *Proceedings of the National Academy of Sciences*, 201617657.

Chase, R. A. (1965). An information-flow model of the organization of motor activity. I: Transduction, transmission and central control of sensory information. *The Journal of Nervous and Mental Disease, 140*, 239–251.

Chen, J. L., Penhune, V. B., & Zatorre, R. J. (2008). Listening to musical rhythms recruits motor regions of the brain. *Cerebral Cortex, 18*, 2844–2854.

Chen, J. L., Rae, C., & Watkins, K. E. (2012). Learning to play a melody: An fMRI study examining the formation of auditory-motor associations. *NeuroImage, 59*, 1200–1208.

Chen, J. L., Zatorre, R. J., & Penhune, V. B. (2006). Interactions between auditory and dorsal premotor cortex during synchornization to musical rhythms. *NeuroImage, 32*, 1771–1781.

Cho, Y. S., Bae, G. Y., & Proctor, R. W. (2012). Referential coding contributes to the horizontal SMARC effect. *Journal of Experimental Psychology: Human Perception and Performance, 38*, 726.

Clarke, E. F. (1987). Categorical rhythm perception: An ecological perspective. In A. Gabrielsson (Ed.), *Action and perception in rhythm and music* (pp. 19–33). Stockholm: Royal Swedish Academy of Music.

Cook, N. (1998). *Music: A very short introduction*. Oxford: Oxford University Press.

Çorlu, M., Maes, P.-J., Muller, C., Kochman, K., & Leman, M. (2015). The impact of cognitive load on operatic singers' timing performance. *Frontiers in Psychology, 6*, 429.

Couchman, J. J., Beasley, R., & Pfordresher, P. Q. (2012). Auditory feedback, self-attribution, and the experience of agency in sequence production. *Consciousness and Cognition, 21*, 186–203.

Craik, F. I. M., & Lockhart, R. S. (1972). Levles of processing: A framework for memory research. *Journal of Verbal Learning and Verbal Behavior, 11*, 671–684.

D'Ausilio, A., Altenmüller, E., Olivetti Belardinelli, M., & Lotze, M. (2006). Cross-modal plasticity of the motor cortex while listening to a rehearsed musical piece. *European Journal of Neuroscience, 24*, 955–958.

Dalla Bella, S., & Berkowska, M. (2009). Singing proficiency in the majority: Normality and "phenotypes" of poor singing. *Annals of the New York Academy of Sciences, 1169*, 99–107.

Dalla Bella, S., Bialunska, A., & Sowinski, P. (2013). Why movement is captured by music, but less by speech: Role of temporal regularity. *PLoS One, 8*, e71945.

Dalla Bella, S., Giguère, J. F., & Peretz, I. (2007). Singing proficiency in the general population. *Journal of the Acoustical Society of America, 121*, 1182–1189.

Dalla Bella, S., Giguère, J. F., & Peretz, I. (2009). Singing in congenital amusia. *Journal of the Acoustical Society of America, 126*, 414–424.

Dalla Bella, S., & Palmer, C. (2011). Tempo effects of timing, tone intensity and finger movements in piano performance. *PLoS One, 6*, e20518.

Davidson, J. W. (1993). Visual perception of performance manner in the movements of solo musicians. *Psychology of Music, 21*, 103–113.

De Souza, J. (2017). Music at hand: Instruments, bodies, and cognition. Oxford University Press.

Dean, R. T., & Dunsmuir, W. T. M. (2016). Dangers and uses of cross-correlation in analyzing time series in perception, performance, movement, and neuroscience: The importance of constructing transfer function autoregressive models. *Behavior Research Methods, 48*, 783–802.

Dehaene, S., Bossini, S., & Giraux, P. (1993). The mental representation of parity and number magnitude. *Journal of Experimental Psychology: General, 122*, 371–396.

Dell, G. S., Burger, L. K., & Svec, W. R. (1997). Language production and serial order: A functional analysis and a model. *Psychological Review, 104*, 123–147.

Demos, A. P., Carter, D. J., Wanderley, M. M., & Palmer, C. (2017). The unresponsive partner: Roles of social status, auditoryfeedback, and animacy in coordination of joint music performance. *Frontiers in Psychology, 8*.

Demos, A. P., Chaffin, R., Begosh, K. T., Daniels, J. R., & Marsh, K. L. (2012). Rocking to the beat: Effects of music and partner's movements on spontaneous interpersonal coordination. *Journal of Experimental Psychology: General, 141*, 49–53.

Deutsch, D. (1970). Tones and numbers: Specificity of interference in immediate memory. *Science, 168*, 1604–1605.

Deutsch, D., Henthorn, T., & Lapidis, R. (2011). Illusory transformation from speech to song. *Journal of the Acoustical Society of America, 129*, 2245–2252.

Diehl, R. L., Lotto, A. J., & Holt, L. L. (2004). Speech perception. *Annual Review of Psychology, 55*, 149–179.

Dolk, T., Hommel, B., Colzato, L. S., Schütz-Bosbach, S., Prinz, W., & Liepelt, R. (2014). The joint Simon effect: A review and theoretical integration. *Frontiers in Psychology, 5*.

Donchin, E., & Coles, M. G. (1988). Is the P300 component a manifestation of context updating? *Behavioral and Brain Sciences, 11*, 357–374.

Donnay, G. F., Rankin, S. K., Lopez-Gonzalez, M., Jiradejvong, P., & Limb, C. J. (2014). Neural substrates of interactive musical improvisation: An fMRI study of 'trading fours' in jazz. *PLoS One, 9*, e88665.

Douglas, K. M., & Bilkey, D. K. (2007). Amusia is associated with deficits in spatial processing. *Nature Neuroscience, 10*, 915.

Dowling, W. J. (1978). Scale and contour: Two components of a theory of memory for melodies. *Psychological Review, 85*, 341–354.

Dowling, W. J., & Bartlett, J. C. (1981). The importance of interval information in long-term memory for melodies. *Psychomusicology, 1*, 30–49.

Dowling, W. J., & Fujitani, D. S. (1971). Contour, interval, and pitch recognition in memory for melodies. *Journal of the Acoustical Society of America, 49*, 524–531.

Drake, C., Jones, M. R., & Baruch, C. (2000). The development of rhythmic attending in auditory sequences: Attunement, referent period, focal attending. *Cognition, 77*, 251–288.

Drake, C., & Palmer, C. (2000). Skill acquisition in music performance: Relations between planning and temporal control. *Cognition, 74*, 1–32.

Drost, U. C., Rieger, M., Brass, M., Gunter, T. C., & Prinz, W. (2005). Action–effect coupling in pianists. *Psychological Research, 69*, 233–241.

Drost, U. C., Rieger, M., & Prinz, W. (2007). Instrument specificity in experienced musicians. *Quarterly Journal of Experimental Psychology, 60*, 527–533.

Eitan, Z., & Timmers, R. (2010). Beethoven's last piano sonata and those who follow crocodiles: Cross-domain mappings of auditory pitch in a musical context. *Cognition, 114*, 405–422.

Ellis, R. J., & Jones, M. R. (2009). The role of accent salience and joint accent structure in meter perception. *Journal of Experimental Psychology: Human Perception and Performance, 35*, 264–280.

Engel, A., Bangert, M., Horbank, D., Hijmans, B. S., Wilkens, K., Keller, P. E., et al. (2012). Learning piano melodies in visuo-motor or audio-motor training conditions and the neural correlates of their cross-modal transfer. *NeuroImage, 63*, 966–978.

Engel, K. C., Flanders, M., & Soechting, J. F. (1997). Anticipatory and sequential motor control in piano playing. *Experimental Brain Research, 113*, 189.

Erdemir, A., & Rieser, J. J. (2016). Singing without hearing: The use of auditory and motor information when singers, instrumentalists, and nonmusicians sing a familiar tune. *Music Perception, 33*, 546–560.

Ericsson, K. A., Krampe, R. T., & Tesch-Römer, C. (1993). The role of deliberate practice in the acquisition of expert performance. *Psychological Review, 100*, 363–406.

Eriksen, B. A., & Eriksen, C. W. (1974). Effects of noise letters upon the identification of a target letter in a nonsearch task. *Perception & Psychophysics, 16*, 143–149.

Essens, P. J., & Povel, D. J. (1985). Metrical and nonmetrical representations of temporal patterns. *Perception & Psychophysics, 37*, 1–7.

Fairbanks, G. (1954). Systematic research in experimental phonetics: 1. A theory of the speech mechanism as a servosystem. *Journal of Speech & Hearing Disorders, 19*, 133–139.

Fairbanks, G., & Guttman, N. (1958). Effects of delayed auditory feedback upon articulation. *Journal of Speech and Hearing Research, 1*, 333–346.

Fairhurst, M. T., Janata, P., & Keller, P. E. (2013). Being and feeling in sync with an adaptive virtual partner: Brain mechanisms underlying dynamic cooperativity. *Cerebral Cortex, 23*, 2592–2600.

Fairhurst, M. T., Janata, P., & Keller, P. E. (2014). Leading the follower: An fMRI investigation of dynamic cooperativity and leader–follower strategies in synchronization with an adaptive virtual partner. *NeuroImage, 84*, 688–697.

Falk, S., Rathcke, T., & Dalla Bella, S. (2014). When speech sounds like music. *Journal of Experimental Psychology: Human Perception and Performance, 40*, 1491–1506.

Finney, S. A. (1997). Auditory feedback and musical keyboard performance. *Music Perception, 15*, 153–174.

Finney, S. A. (1999). Disruptive effects of delayed auditory feedback on motor sequencing *(unpublished Ph.D. thesis)*. Providence: Brown University.

Finney, S. A. (2001a). FTAP: A Linux-based program for tapping and music experiments. *Behavior Research Methods Instruments & Computers, 33*, 65–72.

Finney, S. A. (2001b). Real-time data collection in Linux: A case study. *Behavior Research Methods, Instruments, and Computers, 33*, 167–173.

Finney, S. A. (2016). *In defense of Linux, USB, and MIDI systems for sensorimotor experiments: A response to Schultz and van Vugt (2015)*. Retrieved from: <http://www.sfinney.com/images/pdfs/sf/finney2016a.pdf> 27.08.18.

Finney, S. A., & Palmer, C. (2003). Auditory feedback and memory for music performance: Sound evidence for an encoding effect. *Memory & Cognition, 31*, 51–64.

Finney, S. A., & Warren, W. H. (2002). Delayed auditory feedback and rhythmic tapping: Evidence for a critical interval shift. *Perception & Psychophysics, 64*, 896–908.

Fischinger, T. (2011). An integrative dual route model of rhythm perception and production. *Musicae Scientiae, 15*, 97–105.

Fitts, P. M., & Seeger, C. M. (1953). S–R compatibility: Spatial characteristics of stimulus and response codes. *Journal of Experimental Psychology, 46*, 199–210.

Flach, R. (2005). The transition from synchronization to continuation. *Human Movement Science, 24*, 465–483.

Fodor, J. (1983). *Modularity of mind*. Cambridge, MA: MIT Press.

Foster, N. E. V., Halpern, A. R., & Zatorre, R. J. (2013). Common parietal activation in musical mental transformations across pitch and time. *NeuroImage, 75*, 27–35.

Fraisse, P. (1982). Rhythm and tempo. In D. Deustch (Ed.), *The psychology of music* (pp. 149–180). San Diego, CA: Academic Press.

Fraisse, P., & Vouillaume, C. (1971). Les repères due sujet dans la synchronization et dans la pseudo-synchronisation [The reference points of the subject during synchronization and during pseudo-synchronization]. *L'Année Psychologique, 71*, 359–369.

Franken, M. K., Acheson, D. J., McQueen, J. M., Hagoort, P., & Eisner, F. (2018). Opposing and following responses in sensorimotor speech control: Why responses go both ways. *Psychonomic Bulletin & Review, 25*, 1458–1467.

Furuya, S., & Soechting, J. (2010). Role of auditory feedback in the control of successive keystrokes during piano playing. *Experimental Brain Research, 204*, 223–237.

Gabrielsson, A. (1999). The performance of music. In D. Deutsch (Ed.), *The psychology of music* (Vol. 2, pp. 501–602). San Diego, CA: Academic Press.

Galantucci, B., Fowler, C. A., & Turvey, M. T. (2006). The motor theory of speech perception reviewed. *Psychonomic Bulletin & Review, 13*, 361–377.

Gallese, V., Gernsbacher, M. A., Heyes, C., Hickok, G., & Iacoboni, M. (2011). Mirror neuron forum. *Perspectives on Psychological Science, 6*, 369–407.

Gates, A., Bradshaw, J., & Nettleton, N. (1974). Effect of different delayed auditory feedback intervals on a music performance task. *Perception & Psychophysics, 14*, 21–25.

Gates, A., & Bradshaw, J. L. (1974). Effects of auditory feedback on a musical performance task. *Perception & Psychophysics, 16*, 105–109.

Gehring, W. J., Goss, B., Coles, M. G. H., Meyer, D. E., & Donchin, E. (1993). A neural system for error detection and compensation. *Psychological Science, 4*, 385–390.

Gibson, J. J. (1979). *The ecological approach to visual perception.* Boston: Houghton Mifflin.

Goebl, W., & Palmer, C. (2008). Tactile feedback and timing accuracy in piano performance. *Experimental Brain Research, 186*, 471–479.

Goebl, W., & Palmer, C. (2009). Synchronization of timing and motion among performing musicians. *Music Perception, 26*, 427–438.

Goldiamond, I., Atkinson, C. J., & Bilger, R. C. (1962). Stabilization of behavior and prolonged exposure to delayed auditory feedback. *Science, 135*, 437–438.

Gordon, E. E. (1975). *Learning theory, patterns, and music.* Buffalo, NY: Tometic Associates.

Greenspon, E. B., Pfordresher, P. Q., & Halpern, A. R. (2017). Pitch imitation ability in mental transformations of melodies. *Music Perception: An Interdisciplinary Journal, 34*, 585–604.

Guenther, F. H. (1994). A neural network model of speech acquisition and motor equivalent speech production. *Biological Cybernetics, 72*, 43–53.

Guenther, F. H. (1995). Speech sound acquisition, coarticulation, and rate effects in a neural network model of speech production. *Psychological Review, 102*, 594–621.

Guenther, F. H. (2006). Cortical interactions underlying the production of speech sounds. *Journal of Communication Disorders, 39*, 350–365.

Guenther, F. H. (2016). *Neural control of speech.* MIT Press.

Guenther, F. H., & Vladusich, T. (2012). A neural theory of speech acquisition and production. *Journal of Neurolinguistics, 25*, 408–422.

Hadley, L. V., Novembre, G., Keller, P. E., & Pickering, M. J. (2015). Causal role of motor simulation in turn-taking behavior. *The Journal of Neuroscience, 35*, 16516–16520.

Haegens, S., & Zion Golumbic, E. (2018). Rhythmic facilitation of sensory processing: A review. *Neuroscience and Biobehavioral Reviews, 86*, 150–165.

Haken, H., Kelso, J. A. S., & Bunz, H. (1985). A theoretical model of phase transitions in human hand movements. *Biological Cybernetics, 51*, 347–356.

Halpern, A. R. (1988). Mental scanning in auditory imagery for songs. *Journal of Experimental Psychology: Learning, Memory, and Cognition, 14*, 434–443.

Hashimoto, Y., & Sakai, K. L. (2003). Brain activations during conscious self-monitoring of speech production with delayed auditory feedback: An fMRI study. *Human Brain Mapping, 20*, 22–28.

Haueisen, J., & Knösche, T. R. (2001). Involuntary motor activity in pianists evoked by music perception. *Journal of Cognitive Neuroscience, 13*, 786–792.

Havlicek, L. (1968). Effects of delayed auditory feedback on musical performance. *Journal of Research in Music Education, 16*, 308–318.

Hebb, D. G. (1949). *The organization of behavior*. New York, NY, USA: Wiely and Sons.

Hébert, S., & Peretz, I. (1997). Recognition of music in long-term memory: Are melodic and temporal patterns equal partners?. *Memory & Cognition, 25*, 518–533.

Heffner, C. C., & Slevc, L. R. (2015). Prosodic structure as a parallel to musical structure. *Frontiers in Psychology, 6*, 1962.

Herholz, S. C., Coffey, E. B. J., Pantev, C., & Zatorre, R. J. (2016). Dissociation of neural networks for predisposition and for training-related plasticity in auditory-motor learning. *Cerebral Cortex, 26*, 3125–3134.

Herholz, S. C., & Zatorre, R. J. (2012). Musical training as a framework for brain plasticity: Behavior, function, and structure. *Neuron, 76*, 486–502.

Herrojo Ruiz, M., Maess, B., Altenmüller, E., Curio, G., & Nikulin, V. V. (2017). Cingulate and cerebellar beta oscillations are engaged in the acquisition of auditory-motor sequences. *Human Brain Mapping, 38*, 5161–5179.

Herrojo-Ruiz, M. H., Jabusch, H. C., & Altenmüller, E. (2009). Detecting wrong notes in advance: Neuronal correlates of error monitoring in pianists. *Cerebral Cortex, 19*, 2625–2639.

Hickok, G., Buchsbaum, B., Humphries, C., & Muftuler, T. (2003). Auditory-motor interaction revealed by fMRI: Speech, music and working memory in area Spt. *Journal of Cognitive Neuroscience, 15*, 673–682.

Hickok, G., Houde, J. F., & Rong, F. (2011). Sensorimotor integration in speech processing: Computational basis and neural organization. *Neuron, 69*, 407–422.

Hickok, G., & Poeppel, D. (2007). The cortical organization of speech processing. *Nature Reviews Neuroscience, 8*, 393–402.

Highben, Z., & Palmer, C. (2004). Effects of auditory and motor mental practice in memorized piano performance. *Bulletin of the Council for Research in Music Education, 159*, 58–65.

Hommel, B. (2009). Action control according to TEC (theory of event coding). *Psychological Research, 73*, 512–526.

Hommel, B. (2011). The Simon effect as tool and heuristic. *Acta Psychologica, 136*, 189–202.

Hommel, B. (2015). The theory of event coding (TEC) as embodied-cognition framework. *Frontiers in Psychology, 6*.

Hommel, B., Müsseler, J., Aschersleben, G., & Prinz, W. (2001). The theory of event coding (TEC): A framework for perception and action planning. *Behavioral and Brain Sciences, 24*, 849–937.

Houde, J. F., & Jordan, M. I. (1998). Sensorimotor adaptation in speech production, February *Science, 279*, 1213–1216.

Houde, J. F., & Jordan, M. I. (2002). Sensorimotor adaptation of speech I: Compensation and adaptation. *Journal of Speech, Language, and Hearing Research, 45*, 295–310.

Houde, J. F., & Nagarajan, S. S. (2011). Speech production as state feedback control. *Frontiers in Human Neuroscience, 5*, 82.

Howell, P. (1983). The effect of delaying auditory feedback of selected components of the speech signal. *Perception & Psychophysics, 34*, 387–396.

Howell, P. (2004a). Effects of delayed auditory feedback and frequency-shifted feedback on speech control and some potentials for future development of prosthetic aids for stammering. *Stammering Research, 1*, 34–43.

Howell, P. (2004b). Assessment of some contemporary theories of stuttering that apply to spontaneous speech. *Contemporary Issues in Communicative Sciences and Disorders, 39*, 122–139.

Howell, P. (2011a). Listen to the lessons of The King's Speech. *Nature, 470*, 7.

Howell, P. (2011b). *Recovery from suttering*. London: Psychology Press.

Howell, P., & Archer, A. (1984). Susceptibility to the effects of delayed auditory feedback. *Perception & Psychophysics, 36*, 296–302.

Howell, P., & Au-Yeung, J. (2002). The EXPLAN theory of fluency control applied to the diagnosis of stuttering. In E. Fava (Ed.), *Pathology and therapy of speech disorders* (pp. 75–94). Amsterdam: John Benjamins.

Howell, P. (2001). A model of timing interference to speech control in normal and altered listening conditions applied to the treatment of stuttering. In B. Maassen, W. Hulsijn, R. Kent, H. F. M. Peters, & P. H. M. M. van-Lieshout (Eds.), *Speech motor control in normal and disordered speech* (pp. 291–294). Nijmegen: Uttgeverij Vantilt: Nijmegen.

Howell, P. (2002). The EXPLAN theory of fluency control applied to the treatment of stuttering by altered feedback and operant procedures. In E. Fave (Ed.), *Pathology and therapy of speech disorders*. Amsterdam: John Benjamins.

Howell, P., & Powell, D. J. (1984). Hearing your voice through bone and air: Implications for explanations of stuttering behavior from studies of normal speakers. *Journal of Fluency Disorders, 9*, 247–263.

Howell, P., Powell, D. J., & Khan, I. (1983). Amplitude contour of the delayed signal and interference in delayed auditory feedback tasks. *Journal of Experimental Psychology: Human Perception and Performance, 9*, 772–784.

Howell, P., & Sackin, S. (2002). Timing interference to speech in altered listening conditions. *Journal of the Acoustical Society of America, 111*, 2842–2852.

Howell, P., West, R., & Cross, I. (1991). *Representing musical structure*. San Diego, CA: Academic Press.

Huggins, A. W. F. (1968). Delayed auditory feedback and the temporal properties of the speech material. *Zeitschrift für Phonetik, Spachwissenschaft, und Kommunikations-Forschung, 21*, 53–60.

Huron, D., & Royal, M. (1996). What is melodic accent? Converging evidence from musical practice. *Music Perception, 13*, 489–516.

Hutchins, S., Gosselin, N., & Peretz, I. (2010). Identification of changes along a continuum of speech intonation is impaired in congenital amusia. *Frontiers in Psychology, 1*, 236.

Hutchins, S., & Moreno, S. (2013). The linked dual representation model of vocal perception and production. *Frontiers in Psychology, 4*, 825.

Hutchins, S., & Peretz, I. (2012a). A frog in your throat or in your ear? Searching for the causes of poor singing. *Journal of Experimental Psychology: General, 141*, 76–97.

Hutchins, S., & Peretz, I. (2012b). Amusics can imitatie what they cannot discriminate. *Brain & Language, 123*, 234–239.

Hutchins, S., & Peretz, I. (2013). Vocal pitch shift in congenital amusia (pitch deafness). *Brain & Language, 125*, 106–117.

Hutchins, S., Roquet, C., & Peretz, I. (2012). The vocal generosity effect: How bad can your singing be? *Music Perception, 30*, 147–159.

Hyde, K. L., & Peretz, I. (2004). Brains that are out of tune but in time. *Psychological Science, 15*, 356–360.

Ivry, R. B., & Spencer, R. M. C. (2004). The neural representation of time. *Current Opnion in Biology, 14*, 225–232.

Jack, R. H., Mehrabi, A., Stockman, T., & McPherson, A. (2018). Action–sound latency and the perceived quality of digital musical instruments. *Comparing Professional Percussionists and Amateur Musicians, 36*, 109–128.

James, W. (1890). *The principles of psychology* (Vol. 2). New York: Holt.

Janata, P., Tomic, S. T., & Haberman, J. M. (2012). Sensorimotor coupling in music and the psychology of the groove. *Journal of Experimental Psychology: General, 141*, 54–75.

Jebb, A. T., & Pfordresher, P. Q. (2016). Exploring perception–action relations in music production: The asymmetric effect of tonal class. *Journal of Experimental Psychology: Human Perception and Performance, 42*, 658–670.

Jones, J. A., & Keough, D. (2008). Auditory-motor mapping for pitch control in singers and nonsingers. *Experimental Brain Research, 190*, 279–287.

Jones, J. A., & Munhall, K. G. (2000). Perceptual calibration of F0 production: Evidence from feedback perturbation. *Journal of the Acoustical Society of America, 108*.

Jones, M. R. (1976). Time, our lost dimension—toward a new theory of perception, attention, and memory. *Psychological Review, 83*, 323–355.

Jones, M. R. (1987). Dynamic pattern structure in music—recent theory and research. *Perception & Psychophysics, 41*, 621–634.

Jones, M. R., & Boltz, M. G. (1989). Dynamic attending and responses to time. *Psychological review, 96*, 459–491.

Jones, M. R., & Pfordresher, P. Q. (1997). Tracking musical patterns using joint accent structure. *Canadian Journal of Experimental Psychology—Revue Canadienne De Psychologie Experimentale, 51*, 271–291.

Jordan, M. I., & Rumelhart, D. E. (1992). Forward models: Supervised learning with a distal teacher. *Cognitive Science, 16*, 307–354.

Kajihara, T., Verdonschot, R., Sparks, J., & Stewart, L. (2013). Action–perception coupling in violinists. *Frontiers in Human Neuroscience, 7*, 349.

Kalinowski, J., & Saltuklaroglu, T. (2003). Speaking with a mirror: Engagement of mirror neurons via choral speech and its derivatives induces stuttering inhibition. *Medical Hypotheses, 60*, 538–543.

Kalmus, H., Denes, P., & Fry, D. B. (1955). Effect of delayed acoustic feed-back on some non-vocal activities. *Nature, 175*, 1078–1078.

Katz, D. I., & Lackner, J. R. (1977). Adaptation to delayed auditory feedback. *Perception & Psychophysics, 22*, 476–486.

Kawato, M. (1999). Internal models for motor control and trajectory planning. *Current Opinion in Biology, 9*, 718–727.

Kawato, M., Furukawa, K., & Suzuki, R. (1987). A hierarchical neural-network model for control and learning of voluntary movement. *Biological Cybernetics, 57*, 169–185.

Keele, S. W., Ivry, R. B., Mayr, U., Hazeltine, E., & Heuer, H. (2003). The cognitive and neural architecture of sequence representation. *Psychological Review, 110*, 316–339.

Keller, P. E. (2001). Attentional resource allocation in musical ensemble performance. *Psychology of Music, 29*.

Keller, P. E. (2012). Mental imagery in music performance: Underlying mechanisms and potential benefits. *Annals of the New York Academy of Sciences, 1252*, 206–213.

Keller, P. E. (2008). Joint action in music performance. In F. Morganti, A. Carassa, & G. Riva (Eds.), *Enacting intersubjectivity: A cognitive and social perspective to the study of interactions* (pp. 205–221). Amsterdam: IOS Press.

Keller, P. E., & Appel, M. (2010). Individual differences, auditory imagery, and the coordination of body movements and sounds in musical ensembles. *Music Perception: An Interdisciplinary Journal, 28*, 27–46.

Keller, P. E., & Koch, I. (2008). Action planning in sequential skills: Relations to music performance. *The Quarterly Journal of Experimental Psychology, 61*, 275–291.

Keller, P. E., Novembre, G., & Loehr, J. D. (2016). Musical ensemble performance: Representing self, other, and joint action outcomes. In S. S. Obhi, & E. S. Cross (Eds.), *Shared representations: Sensorimotor foundations of social life* (pp. 280–310). Cambridge, UK: Cambridge University Press.

Kelso, J. A. S. (1995). *Dynamical patterns: The self-organization of brain and behavior*. Cambridge, MA: MIT Press.

Keough, D., & Jones, J. A. (2011). Contextual cuing contributes to the independent modification of multiple internal models for vocal control. *Journal of Neurophysiology, 105*, 2448–2456.

Klapp, S. T., & Jagacinski, R. J. (2011). Gestalt principles in the control of motor action. *Psychological Bulletin, 137*, 443–462.

Kleber, B., Friberg, A., Zeitouni, A., & Zatorre, R. (2017). Experience-dependent modulation of right anterior insula and sensorimotor regions as a function of noise-masked auditory feedback in singers and nonsingers. *NeuroImage, 147*, 97–110.

Kleber, B., Zeitouni, A. G., Friberg, A., & Zatorre, R. J. (2013). Experience-dependent modulation of feedback integration during singing: Role of the right anterior insula. *The Journal of Neuroscience, 33*, 6070–6080.

Koelsch, S., Gunter, T. C., von Cramon, Y., Zysset, S., Lohmann, G., & Friederici, A. D. (2002). Bach speaks: A cortical "language-network" serves the processing of music. *NeuroImage, 17*, 956–966.

Körding, K. P., & Wolpert, D. M. (2004). Bayesian integration in sensorimotor learning. *Nature, 427*, 244.

Körding, K. P., & Wolpert, D. M. (2006). Bayesian decision theory in sensorimotor control. *Trends in Cognitive Sciences, 10*, 319–326.

Krampe, R. T., Mayr, U., & Kliegl, R. (2005). Timing, sequencing, and executive control in repetitive movement production. *Journal of Experimental Psychology: Human Perception and Performance, 31*, 379–397.

Krumhansl, C. L. (1990). *Cognitive foundations of musical pitch*. New York: Oxford University Press.

Krumhansl, C. L. (2000). Rhythm and pitch in music cognition. *Psychological Bulletin, 126*, 159–179.

Krumhansl, C. L., & Cuddy, L. L. (2010). A theory of tonal hierarchies in music. In M. R. Jones, R. R. Fray, & A. N. Popper (Eds.), *Music perception* (pp. 51–88). New York: Springer.

Kulpa, J. D., & Pfordresher, P. Q. (2013). Effects of delayed visual and auditory feedback on sequence production. *Experimental Brain Research, 224*, 69–77.

Kunert, R., Willems, R. M., Casasanto, D., Patel, A. D., & Hagoort, P. (2015). Music and language syntax interact in Broca's area: An fMRI study. *PLoS One, 10*, e0141069.

Lachmair, M., Cress, U., Fissler, T., Kurek, S., Leininger, J., & Nuerk, H. -C. (in press). Music–space associations are grounded, embodied and situated: Examination of cello experts and non-musicians in a standard tone discrimination task. *Psychological Research*.

Laeng, B., & Park, A. (1999). Handedness effect on playing a reversed or normal keyboard. *Laterality, 4*, 363–377.

Lahav, A., Saltzman, E., & Schlaug, G. (2007). Action representation of sound: Audiomotor recognition network while listening to newly acquired actions. *The Journal of Neuroscience, 27*, 308–314.

Large, E. W. (1993). Dynamic programming for the analysis of serial behaviors. *Behavior Research Methods, Instruments, & Computers, 25*, 238–241.

Large, E. W. (2010). Neurodynamics of music. In M. R. Jones, R. R. Fay, & A. N. Popper (Eds.), *Music perception* (pp. 201–231). New York: Springer.

Large, E. W., & Jones, M. R. (1999). The dynamics of attending: How people track time-varying events. *Psychological Review, 106*, 119–159.

Lashley, K. (1951). The problem of serial order in behavior. In L. A. Jeffress (Ed.), *Cerebral Mechanisms in Behavior* (pp. 112–136). New York: Wiley.

Larrouy-Maestri, P., & Pfordresher, P. Q. (2018). Pitch perception in music: Do scoops matter? *Journal of Experimental Psychology: Human Perception and Performance, 44*, 1523–1541.

Leaver, A. M., Van Lare, J., Zielinski, B., Halpern, A. R., & Rauschecker, J. (2009). Brain activation during anticipation of sound sequences. *The Journal of Neuroscience, 29*, 2477–2485.

Lee, B. S. (1950). Effects of delayed speech feedback. *Journal of the Acoustical Society of America, 22*, 824–826.

Lee, B. S. (1951). Artificial stutter. *Journal of Speech and Hearing Disorders, 16*, 53–55.

Lega, C., Stephan, M. A., Zatorre, R. J., & Penhune, V. (2016). Testing the role of dorsal premotor cortex in auditory-motor association learning using transcranical magnetic stimulation (TMS). *PLoS One, 11*, e0163380.

Leman, M. (2008). *Embodied music cognition and mediation technology*. Cambridge, MA: MIT Press.

Leman, M., & Maes, P.-J. (2014). The role of embodiment in the perception of music. *Empirical Musicology Review, 9*, 236–246.

Lerdahl, F., & Jackendoff, R. (1983). *A generative theory of tonal music*. Cambridge, Massachusetts: MIT Press.

Levelt, W. J. (1993). *Speaking: From intention to articulation*. Cambridge, MA: MIT Press.

Levelt, W. J. M. (1983). Monitoring and self-repair in speech. *Cognition, 14*, 41–104.

Liberman, A. M., Cooper, F. S., Shankweiler, D. P., & Studdert-Kennedy, M. (1967). Perception of the speech code. *Psychological Review, 74*, 431.

Liberman, A. M., & Mattingly, I. G. (1985). The motor theory of speech perception revised. *Cognition, 21*, 1–36.

Lidji, P., Kolinsky, R., Lochy, A., & Morais, J. (2007). Spatial associations for musical stimuli: A piano in the head? *Journal of Experimental Psychology: Human Perception and Performance, 33*, 1189–1207.

Lidji, P., Palmer, C., Peretz, I., & Morningstar, M. (2011). Listeners feel the beat: Entrainment to English and French speech rhythms. *Psychonomic Bulletin & Review, 18*, 1035–1041.

Lima, C. F., Lavan, N., Evans, S., Agnew, Z., Halpern, A. R., Shanmugalingam, P., et al. (2015). Feel the noise: Relating individual differences in auditory imagery to the structure and function of sensorimotor systems. *Cerebral Cortex, 25*, 4638–4650.

Lindenberger, U., Li, S. C., Gruber, W., & Müller, V. (2009). Brains swinging in concert: Cortical phase synchronization while playing guitar. *BMC Neuroscience, 10*, 22.

Liu, H., & Larson, C. R. (2007). Effects of perturbation magnitude and voice F0 level on the pitch-shift reflex. *The Journal of the Acoustical Society of America, 122*, 3671–3677.

Livingstone, S. R., Thompson, W. F., & Russo, F. A. (2009). Facial expressions and emotional singing: A study of perception and production with motion capture and electromyography. *Music Perception, 26*, 475–488.

Loehr, J. D., Kourtis, D., Vesper, C., Sebanz, N., & Knoblich, G. (2013). Monitoring individual and joint action outcomes in duet music performance. *Journal of Cognitive Neuroscience, 25*, 1049–1061.

Loehr, J. D., & Palmer, C. (2007). Cognitive and biomechanical influences in pianists' finger tapping. *Experimental Brain Research, 178*, 518–528.

Loehr, J. D., & Palmer, C. (2009). Subdividing the beat: Auditory and motor contributions to synchronization. *Music Perception, 26*, 415–425.

Loehr, J. D., & Palmer, C. (2011). Temporal coordination between performing musicians. *Quarterly Journal of Experimental Psychology, 64*, 2153–2167.

Logothetis, N. K. (2008). What we can do and what we cannot do with fMRI. *Nature, 453*, 869–878.

Lotze, M., Scheler, G., Tan, H. R. M., Braun, C., & Birbaumer, N. (2003). The musician's brain: Functional imaging of amateurs and professionals during performance and imagery. *NeuroImage, 20*, 1817–1829.

Lotze, R. H. (1852). *Medizineische psychologie oder physiologie der seele*. Leipzig, Germany: Weidmann'sche Buchhandlung.

Loui, P., Alsop, D., & Schlaug, G. (2009). Tone deafness: A new disconnection syndrome? *The Journal of Neuroscience, 29*, 10215–10220.

Loui, P., Guenther, F. H., Mathys, C., & Schlaug, G. (2008). Action–perception mismatch in tone-deafness. *Current Biology, 18*, R331–R332.

MacKay, D. G. (1968). Metamorphosis of a critical interval: Age-linked changes in the delay in auditory feedback that produces maximal disruption of speech. *Journal of the Acoustical Society of America, 19*.

Mackay, D. G. (1970). How does language familiarity influence stuttering under delayed auditory feedback? *Perceptual and Motor Skills, 30*, 655–669.

MacKay, D. G. (1987). *The organization of perception and action: A theory for language and other cognitive skills*. New York: Springer-Verlag.

MacKay, D. G., & Burke, B. D. (1972). *Differences between syllable repetition and sentence production under delayed auditory feedback* (unpublished data).

Madison, G., & Merker, B. (2004). Human sensorimotor tracking of continuous subliminal deviations from isochrony. *Neuroscience Letters, 370*, 69–73.

Maes, P.-J. (2016). Sensorimotor grounding of musical embodiment and the role of prediction: A review. *Frontiers in Psychology, 7*.

Maes, P.-J., Leman, M., Palmer, C., & Wanderley, M. (2014). Action-based effects on music perception. *Frontiers in Psychology, 4*.

Maes, P.-J., Wanderley, M. M., & Palmer, C. (2015). The role of working memory in the temporal control of discrete and continuous movements. *Experimental Brain Research, 233*, 263–273.

Maess, B., Koelsch, S., Gunter, T. C., & Friederici, A. D. (2001). Musical syntax is processed in Broca's area: An MEG study. *Nature Neuroscience, 4*, 540–545.

Maidhof, C., Rieger, M., Prinz, W., & Koelsch, S. (2009). Nobody is perfect: ERP effects prior performance errors in musicians indicate fast monitoring processes. *PLoS One, 4*, e5032.

Maidhof, C., Vavatzanidis, N., Prinz, W., Rieger, M., & Koelsch, S. (2010). Processing expectancy violations during music performance and perception: An ERP study. *Journal of Cognitive Neuroscience, 22*, 2401–2413.

Marin, O. S., & Perry, D. W. (1999). Neurological aspects of music perception and performance. In D. Deutsch (Ed.), *The psychology of music* (2nd ed., pp. 653–724). San Diego, CA: Academic Press.

Martin, C. D., Niziolek, C. A., Duñabeitia, J. A., Perez, A., Hernandez, D., Carreiras, M., et al. (2018). Online adaptation to altered auditory feedback is predicted by auditory acuity and not by domain-general executive control resources. *Frontiers in Human Neuroscience, 12*.

Martin, J. G. (1972). Rhythmic (hierarchical) versus serial structure in speech and other behavior. *Psychological Review, 79*, 485–509.

Matheson, A. M. M., & Sakata, J. T. (2015). Relationship between the sequencing and timing of vocal motor elements in birdsong. *PLoS One, 10*, e0143203.

Mathias, B., Gehring, W. J., & Palmer, C. (2017). Auditory N1 reveals planning and monitoring processes during music performance. *Psychophysiology, 54*, 235–247.

Mathias, B., Palmer, C., Perrin, F., & Tillmann, B. (2015). Sensorimotor learning enhances expectations during auditory perception. *Cerebral Cortex, 25*, 2238–2254.

Mathias, B., Pfordresher, P. Q., & Palmer, C. (2015). Context and meter enhance long-range planning in music performance. *Frontiers in Human Neuroscience, 8*.

Max, L., Guenther, F. H., Gracco, V. L., Ghosh, S. S., & Wallace, M. E. (2004). Unstable or insufficiently activated internal models and feedback-biased motor control as sources of dysfluency: A theoretical model of stuttering. *Contemporary Issues in Communicative Sciences and Disorders, 31*, 105–122.

McAuley, J. D. (1995). *Perception of time as phase: Toward an adaptive-oscillator model of rhythmic pattern processing*. Unpublished PhD thesis, Indian University, Bloomington, Indiana.

McAuley, J. D., Jones, M. R., Holub, S., Johnston, H. M., & Miller, N. S. (2006). The time of our lives: Lifespan development of timing and event tracking. *Journal of Experimental Psychology: General, 135*, 348–367.

McNamara, A., Buccino, G., Menz, M. M., Gläscher, J., Wolbers, T., Baumgärtner, A., et al. (2008). Neural dynamics of learning sound–action associations. *PLoS One, 3*, e3845.

Meyer, R. K., & Palmer, C. (2003). Temporal and motor transfer in music performance. *Music Perception, 21*, 81–104.

Miller, G. A. (1956). The magical number seven, plus or minus two: Some limits on our capacity for processing information. *Psychological Review, 63*, 81.

Milton, F., Longmore, C. A., & Wills, A. J. (2008). Processes of overall similarity sorting in free classification. *Journal of Experimental Psychology: Human Perception and Performance, 34*, 676–692.

Mishra, J. (2002). A qualitative analysis of strategies employed in efficient and inefficient memorization. *Bulletin of the Council for Research in Music Education, 152*, 74–86.

Moisse, K., Woodruff, B., & Zak, J. H. L. (November 14, 2011). Gabby Giffords: Finding words through song. *ABC News*. Retrieved from: <https://abcnews.go.com/Health/w_MindBodyNews/gabby-giffords-finding-voice-music-therapy/story?id = 14903987>

Moreno, S., & Bidelman, G. M. (2014). Examining neural plasticity and cognitive benefit through the unique lens of musical training. *Hearing Research, 308*, 84–97.

Mukamel, R., Ekstrom, A. D., Kaplan, J., Iacoboni, M., & Fried, I. (2010). Single-neuron responses in humans during execution and observation of actions. *Current Biology, 20*, 750–756.

Müller, K., Aschersleben, G., Esser, K. H., & Müsseler, J. (2000). Effects of delayed auditory feedback on speech:Just a problem of displaced rhythm? In L. W. P. Desain (Ed.), *Rhythm perception and production* (pp. 249–262). Lisse, the Netherlands: Swets & Zeitlinger.

Mürbe, D., Friedemann, P., Hofmann, G., & Sundberg, J. (2002). Significance of auditory and kinesthetic feedback to singers' pitch control. *Journal of Voice, 16*, 44–51.

Mürbe, D., Friedemann, P., Hofmann, G., & Sundberg, J. (2004). Effects of a professional solo singer education on auditory and kinesthetic feedback—A longitudinal study of singers' pitch control. *Journal of Voice, 18*, 236–241.

Müsseler, J., & Hommel, B. (1997a). Blindness to response–compatible stimuli. *Journal of Experimental Psychology: Human Perception and Performance, 23*, 861–872.

Müsseler, J., & Hommel, B. (1997b). Detecting and identifying response–compatible stimuli. *Psychonomic Bulletin & Review, 4*, 125–129.

Mutschler, I., Schulze-Bonhage, A., Glauche, V., Demandt, E., Speck, O., & Ball, T. (2007). A rapid sound–action association effect in human insular cortex. *PLoS One, 2*, e259.

Norton, A., Zipse, L., Marchina, S., & Schlaug, G. (2009). Melodic intonation therapy. *Annals of the New York Academy of Sciences, 1169*, 431–436.

Novembre, G., Knoblich, G., Dunne, L., & Keller, P. E. (2017). Interpersonal synchrony enhanced through 20 Hz phase-coupled dual brain stimulation. *Social Cognitive and Affective Neuroscience, 12*, 662–670.

Novembre, G., Sammler, D., & Keller, P. E. (2016). Neural alpha oscillations index the balance between self-other integration and segregation in real-time joint action. *Neuropsychologia, 89*, 414–425.

Novembre, G., Ticini, L. F., Schütz-Bosbach, S., & Keller, P. E. (2012). Distinguishing self and other in joint action. Evidence from a musical paradigm. *Cerebral Cortex, 22*, 22894–22903.

Oxenham, A. J. (2012). Pitch perception. *The Journal of Neuroscience, 32*, 13335–13338.

Pa, J., & Hickok, G. (2008). A parietal–temporal sensory-motor integration area for the human vocal tract: Evidence from an fMRI study of skilled musicians. *Neuropsychologia, 46*, 362–368.

Palmer, C. (1997). Music performance. *Annual Review of Psychology, 48*, 115.

Palmer, C., & Krumhansl, C. L. (1987). Independent temporal and pitch structures in determination of musical phrases. *Journal of Experimental Psychology: Human Perception and Performance, 13*, 116–126.

Palmer, C., & Krumhansl, C. L. (1990). Mental representations for musical meter. *Journal of Experimental Psychology: Human Perception & Performance, 16*, 728–741.

Palmer, C., Lidji, P., & Peretz, I. (2014). Losing the beat: Deficits in temporal coordination. *Philosophical Transactions of the Royal Society of London, Series B: Biological Sciences, 369*, 20130405.

Palmer, C., & Meyer, R. K. (2000). Conceptual and motor learning in music performance. *Psychological Science, 11*, 63–68.

Palmer, C., & Pfordresher, P. Q. (2003). Incremental planning in sequence production. *Psychological Review, 110*, 683–712.

Palmer, C., & van de Sande, C. (1993). Units of knowledge in music performance. *Journal of Experimental Psychology: Learning Memory and Cognition, 19*, 457–470.

Palmer, C., & van de Sande, C. (1995). Range of planning in music performance. *Journal of Experimental Psychology: Human Perception and Performance, 21*, 947–962.

Parncutt, R., Sloboda, J. A., Clarke, E. F., Raekallio, M., & Desain, P. (1997). An ergonomic model of keyboard fingering for melodic fragments. *Music Perception: An Interdisciplinary Journal, 14*, 341–382.

Pascual-Leone, A., Nguyet, D., Cohen, L. G., Brasil-Neto, J. P., Cammarota, A., & Hallett, M. (1995). Modulation of muscle responses evoked by transcranial magnetic stimulation during the acquisition of new fine motor skills. *Journal of Neurophysiology, 74*, 1037–1045.

Patel, A. D. (2008). *Music, language, and the brain*. New York: Oxford University Press.

Patel, A. D. (2011). Why would musical training benefit the neural encoding of speech? The OPERA hypothesis. *Frontiers in Psychology, 2*, 142.

Patel, A. D. (2014). Can nonlinguistic musical training change the way the brain processes speech? The expanded OPERA hypothesis. *Hearing Research, 308*, 98–108.

Patel, A. D., & Daniele, J. R. (2003). An empirical comparison of rhythm in language and music. *Cognition, 87*, B35–B45.

Patel, A. D., Iversen, J. R., Wassenaar, M., & Hagoort, P. (2008). Musical syntactic processing in agrammatic Broca's aphasia. *Aphasiology, 22*, 776–789.

Pearson, J., & Kosslyn, S. M. (2015). The heterogeneity of mental representation: Ending the imagery debate. *Proceedings of the National Academy of Sciences, 112*, 10089–10092.

di Pellegrino, G., Fadiga, L., Fogassi, L., Gallese, V., & Rizzolatti, G. (1992). Understanding motor events: A neurophysiological study. *Experimental Brain Research, 91*, 176–180.

Peretz, I. (2008). Musical disorders: From behavior to genes. *Current Directions in Psychological Science, 17*, 329–333.

Peretz, I. (2009). Music, language, and modularity framed in action. *Psychologica Belgica, 49*, 157–175.

Peretz, I. (2016). Neurobiology of congenital amusia. *Trends in Cognitive Sciences, 20*, 857–867.

Peretz, I., & Coltheart, M. (2003). Modularity of music processing. *Nature Neuroscience, 6*, 688–691.

Peretz, I., & Hyde, K. L. (2004). What is specific to music processing? Insights from congenital amusia. *Trends in Cognitive Sciences, 7*, 362–367.

Peretz, I., Vuvan, D. T., Lagrois, M.-E., & Armony, J. L. (2015). Neural overlap in processing music and speech. *Philosophical Transactions of the Royal Society of London, Series B: Biological Sciences, 370*, 20140090.

Pfordresher, P. Q. (2003a). Auditory feedback in music performance: Evidence for a dissociation of sequencing and timing. *Journal of Experimental Psychology: Human Perception and Performance, 29*, 949–964.

Pfordresher, P. Q. (2003b). The role of melodic and rhythmic accents in musical structure. *Music Perception, 20*, 431–464.

Pfordresher, P. Q. (2005). Auditory feedback in music performance: The role of melodic structure and musical skill. *Journal of Experimental Psychology: Human Perception and Performance, 31*, 1331–1345.

Pfordresher, P. Q. (2007). Response/effect congruity and disruption of timing by delayed auditory feedback during sequence production. *Journal of Cognitive Neuroscience* (Supplement), 150.

Pfordresher, P. Q. (2008). Auditory feedback in music performance: The role of transition-based similarity. *Journal of Experimental Psychology: Human Perception and Performance, 34*, 708–725.

Pfordresher, P. Q. (2012). Musical training and the role of auditory feedback during performance. *Annals of the New York Academy of Sciences, 1252*, 171–178.

Pfordresher, P. Q. (2014). "Deafness" effects in detecting alterations to auditory feedback during sequence production. *Psychological Research, 78*, 96–112.

Pfordresher, P. Q., & Beasley, R. T. E. (2014). Making and monitoring errors based on altered auditory feedback. *Frontiers in Psychology, 5*, 914.

Pfordresher, P. Q., & Benitez, B. (2007). Temporal coordination between actions and sound during sequence production. *Human Movement Science, 26*, 742–756.

Pfordresher, P. Q., & Brown, S. (2007). Poor-pitch singing in the absence of "tone deafness". *Music Perception, 25*, 95–115.

Pfordresher, P. Q., & Brown, S. (2009). Enhanced production and perception of musical pitch in tone language speakers. *Attention Perception & Psychophysics, 71*, 1385–1398.

Pfordresher, P. Q., & Brown, S. (2017). Vocal mistuning reveals the origins of musical scales. *Journal of Cognitive Psychology, 29*, 35–52.

Pfordresher, P. Q., & Chow, K. (in press). A cost of musical training? Sensorimotor flexibility in musical sequence learning. *Psychonomic Bulletin & Review.*

Pfordresher, P. Q., & Dalla Bella, S. (2011). Delayed auditory feedback and movement. *Journal of Experimental Psychology: Human Perception and Performance, 37*, 566–579.

Pfordresher, P. Q., & Halpern, A. R. (2013). Auditory imagery and the poor-pitch singer. *Psychonomic Bulletin & Review, 20*, 747–753.

Pfordresher, P. Q., Halpern, A. R., & Greenspon, E. B. (2015). A mechanism for sensorimotor translation in singing: The multi-modal imagery association (MMIA) model. *Music Perception, 32*, 293–302.

Pfordresher, P. Q., Keller, P. E., Koch, I., Palmer, C., & Yildirim, E. (2011). Activation of learned action sequences by auditory feedback. *Psychonomic Bulletin & Review, 18*, 544–549.

Pfordresher, P. Q., & Kobrina, A. (2017). Sensitivity to meter in auditory feedback during music performance. *Psychomusicology: Music, Mind, and Brain, 27*, 54–62.

Pfordresher, P. Q., & Kulpa, J. D. (2011). The dynamics of disruption from altered auditory feedback: Further evidence for a dissociation of sequencing and timing. *Journal of Experimental Psychology: Human Perception and Performance, 37*, 949–967.

Pfordresher, P. Q., & Larrouy-Maestri, P. (2015). On drawing a line through the spectrogram: How do we understand deficits of vocal pitch imitation? *Frontiers in Human Neuroscience, 9*, 271.

Pfordresher, P. Q., & Mantell, J. T. (2009). Singing as a form of vocal imitation: Mechanisms and deficits. In J. Louhivouri, T. Eerola, S. Saarikallio, T. Himberg, & P. S. Eerola (Eds.), Proceedings of the 7th Triennial Conference of European Society for the Cognitive Sciences of Music (pp. 425–430). Jyväskylä, Finland: Jyväskylä University.

Pfordresher, P. Q., & Mantell, J. T. (2012). Effects of altered auditory feedback across effector systems: Production of melodies by keyboard and singing. *Acta Psychologica, 139*, 166–177.

Pfordresher, P. Q., & Mantell, J. T. (2014). Singing with yourself: Evidence for an inverse modeling account of poor-pitch singing. *Cognitive Psychology, 70*, 31–57.

Pfordresher, P. Q., Mantell, J. T., Brown, S., Zivadinov, R., & Cox, J. L. (2014). Brain responses to altered auditory feedback during musical keyboard production: An fMRI study. *Brain Research, 1556*, 28–37.

Pfordresher, P. Q., & Palmer, C. (2002). Effects of delayed auditory feedback on timing of music performance. *Psychological Research/Psychologische Forschung, 66*, 71–79.

Pfordresher, P. Q., & Palmer, C. (2006). Effects of hearing the past, present, or future during music performance. *Perception and Psychophysics, 68*, 362–376.

Pfordresher, P. Q., Palmer, C., & Jungers, M. K. (2007). Speed, accuracy, and serial order in sequence production. *Cognitive Science: A Multidisciplinary Journal, 31*, 1–36.

Phillips-Silver, J., Toiviainen, P., Gosselin, N., Piche, O., Nozaradan, S., Palmer, C., et al. (2011). Born to dance but beat deaf: A new form of congenital amusia. *Neuropsychologia, 49*, 961–969.

Platt, J. R. (1964). Strong inference. *Science, 146*, 347–353.

Pollard, H. (1968). Time delay effects in the operation of a pipe organ. *Acta Acustica United With Acustica, 20*, 189–199.

Poonian, S. K., McFadyen, J., Ogden, J., & Cunnington, R. (2015). Implicit agency in observed actions: Evidence for N1 suppression of tones caused by self-made and observed actions. *Journal of Cognitive Neuroscience, 27*, 752–764.

Port, R. F., & Van Gelder, T. (1995). *Mind as motion: Explorations in the dynamics of cognition.* Cambridge, MA: MIT Press.

Povel, D. J., & Essens, P. (1985). Perception of temporal patterns. *Music Perception, 2*, 411–440.

Pressing, J. (1999). The referential dynamics of cognition and action. *Psychological Review, 106*, 714–747.

Prince, J. B., & Schmuckler, M. A. (2014). The tonal-metric hierarchy: A corpus analysis. *Music Perception, 31*, 245–270.

Prince, J. B., Thompson, W. F., & Schmuckler, M. A. (2009). Pitch and time, tonality and meter: How do musical dimensions combine? *Journal of Experimental Psychology: Human Perception and Performance, 35*, 1598–1617.

Prinz, W. (1997). Perception and action planning. *European Journal of Cognitive Psychology, 9*, 129–154.

Pruitt, T. A., & Pfordresher, P. Q. (2015). The role of auditory feedback in speech and song. *Journal of Experimental Psychology: Human Perception and Performance, 41*, 152–166.

Rabbitt, P. (1978). Detection of errors by skilled typists. *Ergonomics, 21*, 945–958.

Rabbitt, P. M. (1966). Errors and error correction in choice-response tasks. *Journal of Experimental Psychology, 71*, 264–272.

Ragert, M., Schroeder, T., & Keller, P. (2013). Knowing too little or too much: The effects of familiarity with a co-performer's part on interpersonal coordination in musical ensembles. *Frontiers in Psychology, 4*.

Repp, B. H. (1996). The art of inaccuracy: Why pianists' errors are difficult to hear. *Music Perception: An Interdisciplinary Journal, 14*, 161–183.

Repp, B. H. (1999). Effects of auditory feedback deprivation on expressive piano performance. *Music Perception, 16*, 409–438.

Repp, B. H., & Knoblich, G. (2007). Toward a psychophysics of agency: Detecting gain and loss of control over auditory action effects. *Journal of Experimental Psychology: Human Perception and Performance, 33*, 469–482.

Repp, B. H., & Knoblich, G. (2009). Performed or observed keyboard actions affect pianists' judgements of relative pitch. *The Quarterly Journal of Experimental Psychology, 62*, 2156–2170.

Ridderinkhof, K. R., Ullsperger, M., Crone, E. A., & Nieuwenhuis, S. (2004). The role of the medial frontal cortex in cognitive control. *Science, 306*, 443–447.

Robinson, G. M. (1972). The delayed auditory feedback effect is a function of speech rate. *Journal of Experimental Psychology, 95*, 1–5.

Rouder, J. N., Speckman, P. L., Sun, D., Morey, R. D., & Iverson, G. (2009). Bayesian *t* tests for accepting and rejecting the null hypothesis. *Psychonomic Bulletin & Review, 16*, 225–237.

Rusconi, E., Kwan, B., Giordano, B. L., Umilta, C., & Butterworth, B. (2006). Spatial representation of pitch height: The SMARC effect. *Cognition, 99*, 113–129.

Rutkowski, J. (2015). The relationship between children's use of singing voice and singing accuracy. *Music Perception: An Interdisciplinary Journal, 32*, 283–292.

Salamé, Q., & Baddeley, A. (1989). Effects of background music on phonological short-term memory. *The Quarterly Journal of Experimental Psychology Section A, 41*, 107–122.

Sato, A. (2009). Both motor prediction and conceptual congruency between preview and action–effect contribute to explicit judgment of agency. *Cognition, 110*, 74–83.

Sato, A., & Yasuda, A. (2005). Illusion of sense of self-agency: Discrepancy between the predicted and actual sensory consequences of actions modulates the sense of self-agency, but not the sense of self-ownership. *Cognition, 94*, 241–255.

Schendel, Z. A., & Palmer, C. (2007). Suppression effects on musical and verbal memory. *Memory & Cognition, 35*, 640–650.

Schlaug, G., Marchina, S., & Norton, A. (2008). From singing to speaking: Why singing may lead to recovery of expressive language function in patients with Broca's aphasia. *Music Perception: An Interdisciplinary Journal, 25*, 315–323.

Schmidt, R. A. (1975). A schema theory of discrete motor skill learning. *Psychological Review, 82*, 225–260.

Schmidt, R. C., Carello, C., & Turvey, M. T. (1990). Phase transitions and critical fluctuations in the visual coordination of rhythmic movements between people. *Journal of Experimental Psychology: Human Perception and Performance, 16*, 227–247.

Schultz-Coulon, H. J. (1978). The neuromuscular phonatory control system and vocal function. *Acta Oto-Laryngologica, 86*, 142–153.

Scott, S. K. (1998). The point of P-centres. *Psychological Research, 61*, 4.

Sebanz, N., Bekkering, H., & Knoblich, G. (2006). Joint action: Bodies and minds moving together. *Trends in Cognitive Sciences, 10*, 70–76.

Sebanz, N., & Knoblich, G. (2009). Prediction in joint action: What, when, and where. *Topics in Cognitive Science, 1*, 353–367.

Shepard, R. N. (1964). Circularity in judgments of relative pitch. *Journal of the Acoustical Society of America, 36*, 2346.

Shin, Y. K., Proctor, R. W., & Capaldi, E. J. (2010). A review of contemporary ideomotor theory. *Psychological Bulletin, 136*, 943–974.

Siegel, G. M., Fehst, C. A., Garber, S. R., & Pick, J. H. L. (1980). Delayed auditory feedback with children. *Journal of Speech, Language, and Hearing Research, 23*, 802–813.

Simchy-Gross, R., & Margulis, E. H. (2018). The sound-to-music illusion: Repetition can musicalize nonspeech sounds. *Music & Science, 1*, 2059204317731992.

Simon, J. R., & Rudell, A. P. (1967). Auditory S–R compatibility: The effect of an irrelevant cue on information processing. *Journal of Applied Psychology, 51*, 300–304.

Simon, J. R., Sly, P. E., & Vilapakkam, S. (1981). Effect of compatibility of S–R mapping on reactions toward the stimulus source. *Acta Psychologica, 47*, 63–81.

Sloboda, J. A. (1983). The communication of musical metre in piano performance. *The Quarterly Journal of Experimental Psychology Section A, 35*, 377–396.

Sloboda, J. A., Clarke, E. F., Parncutt, R., & Raekallio, M. (1998). Determinants of finger choice in piano sight-reading. *Journal of Experimental Psychology: Human Perception and Performance, 24*, 185–203.

Smith, J. D. (1997). The place of musical novices in music science. *Music Perception: An Interdisciplinary Journal, 14*, 227–262.

Smith, K. U. (1962). *Delayed sensory feedback and behavior*. Philadelphia: W. B. Saunders.

Speer, S. R., Crowder, R. G., & Thomas, L. M. (1993). Prosodic structure and sentence recognition. *Journal of Memory and Language, 32*, 336–358.

Stephan, M. A., Brown, R., Lega, C., & Penhune, V. (2016). Melodic priming of motor sequence performance: The role of the dorsal premotor cortex. *Frontiers in Neuroscience, 10*, 210.

Stewart, L., Verdonschot, R. G., Nasralla, P., & Lanipekun, J. (2013). Action–perception coupling in pianists: Learned mappings or spatial musical association of response codes (SMARC) effect? *The Quarterly Journal of Experimental Psychology, 66*, 37–50.

Stewart, L., Walsh, V., & Frith, U. (2004). Reading music modifies spatial mapping in pianists. *Perception & Psychophysics, 66*, 183.

Stock, A., & Stock, C. (2004). A short history of ideo-motor action. *Psychological Research, 68*, 176–188.

Stoet, G., & Hommel, B. (1999). Action planning and the temporal binding of response codes. *Journal of Experimental Psychology: Human Perception and Performance, 25*.

Subramaniam, K., Kothare, H., Mizuiri, D., Nagarajan, S. S., & Houde, J. F. (2018). Reality monitoring and feedback control of speech production are related through self-agency. *Frontiers in Human Neuroscience, 12*, 82.

Sundberg, J. (1989). *The science of the singing voice*. DeKalb, IL: Northern Illinois University Press.

Sundberg, J., & Bauer-Huppmann, J. (2007). When does a sung tone start? *Journal of Voice, 21*, 285–293.

Takaso, H., Eisner, F., Wise, R. J. S., & Scott, S. K. (2010). The effect of delayed auditory feedback on activity in the temporal lobe while speaking: A positron emission tomography study. *Journal of Speech Language and Hearing Research, 53*, 226–236.

Tan, S.-L., Pfordresher, P. Q., & Harré, R. (2018). *Psychology of music: From sound to significance* (2nd ed.). London: Routledge.

Taylor, J. E., & Witt, J. K. (2015). Listening to music primes space: Pianists, but not novices, simulate heard actions. *Psychological Research, 79*, 175–182.

Thomassen, J. M. (1982). Melodic accent—experiments and a tentative model. *Journal of the Acoustical Society of America, 71*, 1596–1605.

Thompson, W. F., Russo, F. A., & Livingstone, S. R. (2010). Facial expressions of singers influence perceived pitch relations. *Psychonomic Bulletin & Review, 17*, 317–322.

Tierney, A., Dick, F., Deutsch, D., & Sereno, M. (2013). Speech versus song: Multiple pitch-sensitive areas revealed by a naturally occurring musical illusion. *Cerebral Cortex, 23*, 249–254.

Tiffany, W. R., & Hanley, C. N. (1956). Adaptation to delayed sidetone. *Journal of Speech & Hearing Disorders, 21*, 164–172.

Timmers, R., Endo, S., Bradbury, A., & Wing, A. M. (2014). Synchronization and leadership in string quartet performance: A case study of auditory and visual cues. *Frontiers in Psychology, 5*, 645.

Toiviainen, P., Luck, G., & Thompson, M. R. (2010). Embodied meter: Hierarchical eigenmodes in music-induced movement. *Music Perception: An Interdisciplinary Journal, 28*, 59–70.

Trainor, L. J., Shahin, A. J., & Roberts, L. E. (2009). Understanding the benefits of musical training. *Annals of the New York Academy of Sciences, 1169*, 133–142.

Tremblay-Champoux, A., Dalla Bella, S., Phillips-Silver, J., Lebrun, M.-A., & Peretz, I. (2010). Singing proficiency in congenital amusia: Imitation helps. *Cognitive Neuropsychology, 27*, 463–476.

Tulving, E., & Thompson, D. M. (1973). Encoding specificity and retrieval processes in episodic memory. *Psychological Review, 80*.

Vallet, G., Shore, D., & Schutz, M. (2014). Exploring the role of amplitude envelope in duration estimation. *Perception, 43*, 616–630.

Vanden Bosch der Nederlanden, C. M., Hannon, E. E., & Snyder, J. S. (2015). Everyday musical experience is sufficient to perceive the speech-to-song illusion. *Journal of Experimental Psychology: General, 144*, e43–e49.

Vines, B. W., Krumhansl, C. L., Wanderley, M. M., Dalca, I. M., & Levitin, D. J. (2011). Music to my eyes: Cross-modal interactions in the perception of emotions in musical performance. *Cognition, 118*, 157–170.

Vines, B. W., Krumhansl, C. L., Wanderley, M. M., & Levitin, D. J. (2006). Cross-modal interactions in the perception of musical performance. *Cognition, 101*, 80–113.

Vorberg, D., & Wing, A. (1996). Modeling variability and dependence in timing. In H. Heuer, & S. W. Keel (Eds.), *Handbook of perception and action* (Vol. 2, pp. 181–262). London: Academic Press.

Vousden, J. I., & Brown, G. D. A. (1998). To repeat or not to repeat: The time course of response suppression in sequential behaviour. In J. A. Bullinaria, D. W. Glasspool, & G. Houghton (Eds.), *Proceedings of the fourth neural computation and psychology workshop: Connectionist representations*. London: Springer-Verlag.

Van Vugt, F. T., Jabusch, H.-C., & Altenmüller, E. (2012). Fingers phrase music differently: Trial-to-trial variability in piano scale playing and auditory perception reveal motor chunking. *Frontiers in Psychology, 3*, 495.

Vuvan, D. T., Nunes-Silva, M., & Peretz, I. (2015). Meta-analytic evidence for the non-modularity of pitch processing in congenital amusia. *Cortex, 69*, 186–200.

Vuvan, D. T., Paquette, S., Mignault Goulet, G., Royal, I., Felezeu, M., & Peretz, I. (2017). The montreal protocol for identification of amusia. *Behavior Research Methods, 50*, 662–672.

Wan, C. Y., Bazen, L., Baars, R., Libenson, A., Zipse, L., Zuk, J., et al. (2011). Auditory-motor mapping training as an intervention to facilitate speech output in non-verbal children with autism: A proof of concept study. *PLoS One, 6*, e25505.

Wanderley, M. M. (2002). *Quantitative analysis of non-obvious performer gestures. Paper presented at the International Gesture Workshop.*

Ward, W. D., & Burns, E. M. (1978). Singing without auditory feedback. *Journal of Research in Singing, 1*, 24–44.

Watkins, K. E., Patel, N., Davis, S., & Howell, P. (2005). Brain activity during altered auditory feedback: An fMRI study in healthy adolescents. *NeuroImage, 26*(Suppl. 1).

Watson, J. B. (1920). Is thinking merely the action of the language mechanism? *British Journal of Psychology, 11*, 86–104.

West, R., Cross, I., & Howell, P. (1985). *Musical structure and cognition*. San Diego, CA: Academic Press.

Wheeldon, L. R., Meyer, A. S., & Smith, M. (2002). Incrementality. In R. Goldstone (Ed.), *Encyclopedia of cognitive science*. London: Macmillan.

Wegner, D. M., Sparrow, B., & Winerman, L. (2004). Vicarious agency: Experiencing control over the movements of others. *Journal of Personality and Social Psychology, 86*, 838.

Wegner, D. M., & Wheatley, T. (1999). Apparent mental causation: Sources of the experience of will. *American Psychologist, 54*, 480.

Wiener, N. (1948). Cybernetics. *Scientific American, 179*, 14–19.

Wiener, N. (1961). *Cybernetics, or control and communication in the animal and the machine* (Vol. 25). Cambridge, MA: MIT press.

Williamon, A., & Valentine, E. (2002). The role of retrieval structures in memorizing music. *Cognitive Psychology, 44*, 1–32.

Williamson, V. J., Mitchell, T., Hitch, G. J., & Baddeley, A. D. (2010). Musicians' memory for verbal and tonal materials under conditions of irrelevant sound. *Psychology of Music, 38*, 331–350.

Williamson, V. J., Liu, F., Peryer, G., Grierson, M., & Stewart, L. (2012). Perception and action de-coupling in congenital amusia: Sensitivity to task demands. *Neuropsychologia, 50*, 172–180.

Wilson, M. (2002). Six views of embodied cognition. *Psychonomic Bulletin & Review, 9*, 625–636.

Wing, A. M. (1977). Perturbations of auditory feedback delay and the timing of movement. *Journal of Experimental Psychology: Human Perception and Performance, 3*, 175–186.

Wing, A. M., & Kristofferson, A. B. (1973a). Response delays and the timing of discrete motor responses. *Perception & Psychophysics, 14*, 5–12.

Wing, A. M., & Kristofferson, A. B. (1973b). The timing of interresponse intervals. *Perception & Psychophysics, 13*, 455–460.

Woolley, A. G. (2006). *The physical characteristics of mechanical pipe organ actions and how they affect musical performance.* Unpublished PhD thesis, University of Edinburgh.

Wollman, I., Penhune, V., Segado, M., Carpentier, T., & Zatorre, R. J. (in press). Neural network retuning and neural predictors of learning success associated with cello training. *Proceedings of the National Academy of Sciences.*

Wolpert, D. M. (1997). Computational approaches to motor control. *Trends in Cognitive Sciences, 1*, 209–216.

Wolpert, D. M., Ghahramani, Z., & Jordan, M. I. (1995). An internal model for sensorimotor integration. *Science, 269*, 1880–1882.

Wolpert, D. M., & Kawato, M. (1998). Multiple paired forward and inverse models for motor control. *Neural Networks, 11*, 1317–1329.

Wolpert, D. M., Miall, R. C., & Kawato, M. (1998). Internal models in the cerebellum. *Trends in Cognitive Sciences, 2*, 338–347.

Wong, P. C. M., Skoe, E., Russo, N. M., Dees, T., & Kraus, N. (2007). Musical experience shapes human brainstem encoding of linguistic pitch patterns. *Nature Neuroscience, 10*, 420–422.

Zamm, A., Debener, S., Bauer, A. K. R., Bleichner, M. G., Demos, A. P., & Palmer, C. (2018). Amplitude envelope correlations measure synchronous cortical oscillations in performing musicians. *Annals of the New York Academy of Sciences, 1423*, 251–263.

Zamm, A., Pfordresher, P. Q., & Palmer, C. (2015). Temporal coordination in joint music performance: Effects of endogenous rhythms and auditory feedback. *Experimental Brain Research, 233*, 607–615.

Zamm, A., Wellman, C., & Palmer, C. (2016). Endogenous rhythms influence interpersonal synchrony. *Journal of Experimental Psychology: Human Perception and Performance, 42*, 611–616.

Zarate, J. M., Wood, S., & Zatorre, R. J. (2010). Neural networks involved in voluntary and involuntary pitch regulation in exeprienced singers. *Neuropsychologia, 48*, 607–718.

Zarate, J. M., & Zatorre, R. J. (2008). Experience-dependent neural substrates involved in vocal pitch regulation during singing. *NeuroImage, 40*, 1871–1887.

Zatorre, R. J., & Baum, S. R. (2012). Musical melody and speech intonation: Singing a different tune. *PLoS Biology, 10*, e1001372.

Zatorre, R. J., Belin, P., & Penhune, V. B. (2002). Structure and function of auditory cortex: Music and speech. *Trends in Cognitive Sciences, 6*, 37–46.

Zatorre, R. J., Chen, J. L., & Penhune, V. B. (2007). When the brain plays music: Auditory-motor interactions in music perception and production. *Nature Reviews Neuroscience, 8*, 547–558.

Zatorre, R. J., & Halpern, A. R. (2005). Mental concerts: Musical imagery and auditory cortex. *Neuron, 47*, 9–12.

Zatorre, R. J., Halpern, A. R., & Bouffard, M. (2010). Mental reversal of imagined melodies: A role for the posterior parietal cortex. *Journal of Cognitive Neuroscience, 22*, 775–789.

Zheng, Z. Z., MacDonald, E. N., Munhall, K. G., & Johnsrude, I. S. (2011). Perceiving a stranger's voice as being one's own: A 'rubber voice' illusion? *PLoS One, 6*, e18655.

Zipse, L., Norton, A., Marchina, S., & Schlaug, G. (2012). When right is all that is left: Plasticity of right-hemisphere tracts in a young aphasic patient. *Annals of the New York Academy of Sciences, 1252*, 237–245.

Epilogue: What We Have Learned and Where to Go

In this book I have focused on the role of sound (auditory feedback) during music performance with three broad goals in mind: (1) synthesize the state of knowledge at this time, (2) reflect on the broader theoretical and practical implications of these results, and (3) discuss connections of this research with other related topics. In discussing these issues I have developed a theoretical framework based on hierarchical perception/action coordination (H-PAC) that not only follows from the existing literature but also makes predictions that deserve testing. Indeed, although some aspects of how perception and action intersect in music are quite well explored, others are not. In this epilogue I reflect further on these broader issues.

The book began with the Beatles, and I will here draw on them to discuss what we have learned so far from research on the role of auditory feedback.[1] As noted in Chapter 2, the fact that the Beatles could still perform reasonably well with minimal auditory feedback makes sense, given the level of musical expertise in this group along with the fact that they performed songs they either wrote or knew incredibly well. The intensely vivid auditory images each band member would have for the music caused them to effectively experience what they performed even without hearing, similar to Beethoven's ability to compose after becoming deaf (not that I am putting Beethoven and the Beatles in exactly the same category!). A more novice group, or any group rehearsing a song for the first time, by contrast would suffer noticeably from the lack of auditory feedback.

In contrast to the modest effects of removing or masking auditory feedback, even a band as good as the Beatles would have trouble if feedback were presented with a substantial delay as discussed in Chapter 3. This is because we can hardly help synchronizing our actions with perceived rhythmic patterns, even if those perceived patterns ought to be ignored. For the same reason, performing a song with highly complex syncopated rhythms, like "Everybody's Got Something

[1]The historical anecdotes used here are drawn from *The Beatles Anthology* (McCartney, Lennon, Harrison, & Starr, 1995, Chronicle Books).

to Hide Except Me and My Monkey" (a song the Beatles never performed live) would be incredibly difficult, as discussed in Chapter 9.

The Beatles also experimented with alterations of sound content, most famously the use of the Mellotron in "Strawberry Fields Forever" and the reproduction of backwards recording (e.g., the guitar solo in "I'm Only Sleeping"). In Chapter 4, we discussed how the former alteration—influencing timbre but not pitch patterns—would likely cause little disruption (if any) to the performer. By contrast, it would be highly difficult to perform a guitar solo while hearing a temporally reversed copy in real time.

While composing songs, the Beatles would often focus primarily on the melodic aspects, adding words only later and sometimes in a haphazard way. For instance, Paul McCartney did not initially have a lyrical idea for the ballad "Yesterday," but knew that the song should begin with some kind of three-syllable pattern. So, while developing the song, he inserted the phrase "scrambled eggs" as a filler. Even more striking, John Lennon constructed the lyrics for "I Am the Walrus" by writing phrases on scraps of paper, putting them all in a jar, and then retrieving different phrases at random from the jar. These examples illustrate the separability of verbal from melodic information, which was a key theme from the literature summarized in Chapter 7. At the same time, similar principles concerning how perception and action interact hold for both music and speech, which is why it is so important for the rhythmic pattern in song lyrics to cohere well with the rhythmic patterns created by the melody.

And of course, like many rock groups of the 1960s, the Beatles experimented with recreational drug use. One of the key effects of drug use is how higher cognitive processes, in particular the selfregulating effects of executive control, are influenced. The fact that the quality of music did not suffer during their periods of drug use (many think their music improved as a result) reflects the fact that sensorimotor coordination in music does not rely on these functions, as discussed in Chapter 8.

I use this extended example to relate the research summarized in this book to well-known examples from real music making. Although I highlight one particularly well-known rock band, any number of musicians and musical ensembles could be inserted here.

In addition to summarizing research, I also proposed a theoretical framework for how perception and action relate to each other in music performance (and potentially other sequential tasks) based both on existing research and other theoretical ideas from cognitive psychology. The resulting H-PAC model thus borrows many ideas from many sources and puts them all together. It draws on ecological and dynamical systems approaches in proposing that the interactions across perception and action are based on coordination in real time. However, unlike these approaches this model assumes that there are internal, abstract representations that govern these patterns of coordination, based on the fact that many studies suggest that tonal and metrical representations influence responses to auditory feedback during performance. The heart of this model has to do with a coordinative representation that mediates perception and action based on hierarchical divisions between sequencing and timing, borrowed from Node Structure Theory, within a kind of sensorimotor loop, as described in Chapter 5. For the most part, this model is abstract and hypothetical in its expression, although in Chapters 3 and 4, I discussed quantitative implementations that simulate certain effects of altered auditory feedback.

Where do we go from here? Although certain areas of research are fairly well developed (primarily those summarized in the first half of the book), certain topics summarized here are conspicuously underdeveloped. The primary example in my view is the relationship between sensorimotor behavior and higher cognitive processes, discussed in Chapter 8. As noted there, the theoretical extensions of H-PAC that I described are largely hypothetical. Given how critical these higher functions are to cognition in general, it is critical to know more about their interface with lower-level sensorimotor processes. Another issue needing more development has to do with principles of sensorimotor learning discussed in Chapter 7. Although many studies in the past decade have explored how training influences associations between perception and action, we still know little about the kind of constraints on these associations arising from general life experience, and how these constraints interact with acquired musical training. Finally, although a tremendous surge in research on joint performance has arisen recently (Chapter 9), the specific role of auditory information and its relationship to self-feedback is still not well understood.

Despite these areas of potential growth, I thought enough had been done to warrant a new book devoted to the role of sound in music performance from a cognitive science perspective, complementing the musicological treatment of De Souza (2017). For almost two decades I have devoted much of my time to understanding this issue and I hope that my fascination with the topic came through in the chapters presented here. For me, this very specific issue provides a gateway towards a wide range of fascinating issues relating to perception and action, our mind's representation for sequential structures, learning, modularity, and interpersonal coordination.

Author Index

Note: Page numbers followed by "*f*" refer to figures.

G

H

I

J

K

Subject Index

Note: Page numbers followed by "*f*" and "*t*" refer to figures and tables, respectively.

E

F

G

H

FSC
www.fsc.org
MIX
Papier | Fördert
gute Waldnutzung
FSC® C083411

Kolibri 360 GmbH
Ferdinand-Jühlke-Straße 7
99095 Erfurt, Deutschland
produktsicherheit@kolibri360.de